"The lights underneath it flashed red, yellow, blue, sequentially. We all watched it for three or four minutes. It then shot off horizontally. I'll never forget the total silence on our street."

"When I first started reading *Communion*, I remember saying out loud, 'They are not visitors. They are simply Others.' What that means, I really have no idea."

"They peeled back the air like it was the page of a book . . ."

"We found ourselves sitting in our cars, revving the engines, and realizing we didn't know why we were doing this. But we were unable to stop."

"When I first saw your book, I thought the poor author would never know how close the depiction of the being on the cover was to the real thing."

"I am a 48-year-old police detective. Frankly, the books scare the hell out of me. Tell me I am imagining things."

**WHAT YOU ARE READING IS TRUE.
IT IS ACTUAL TESTIMONY FROM
PERSONAL ENCOUNTERS.
AND THERE IS MUCH,
MUCH MORE . . .**

Books by Whitley Strieber

Communion
Transformation
Majestic
The Wolfen
The Hunger
Warday
Nature's End
*Breakthrough**
*The Secret School**

With Anne Strieber

*The Communion Letters**

**Published by* HarperCollins*Publishers*

THE COMMUNION LETTERS

EDITED BY

WHITLEY STRIEBER

AND

ANNE STRIEBER

HarperPrism
A Division of HarperCollins*Publishers*

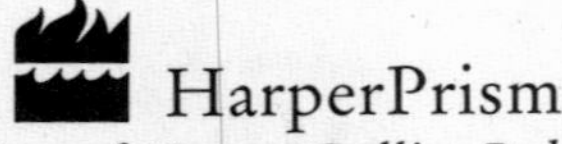

A Division of HarperCollins*Publishers*
10 East 53rd Street, New York, N.Y. 10022-5299

ISBN: 0-06-105368-6

Printed in the United States of America

Cover illustration © 1997 by Donna Diamond

First printing: December 1997

Designed by Lisa Pifher

Library of Congress Cataloging-in-Publication Data
is available from the publisher.

97 98 99 00 ❖ 10 9 8 7 6 5 4 3 2

This book is gratefully dedicated to the nearly two hundred thousand people who have written us describing their own experiences. Without the goodness of soul that shines on every page that you sent us, and the support that you have given us, we could not have endured.

ACKNOWLEDGMENTS

FROM WHITLEY AND ANNE STRIEBER

We would both like to thank the many hundreds of people who had the kindness and courage to agree to allow their letters to us to be excerpted in this volume. To those whose narratives did not appear, please know that it was only from lack of space.

We would especially like to thank our associate, Lorie Barnes, who over the years has typed over thirty thousand of the handwritten letters, and whose care and dedication has made the Communion Foundations letters archive possible.

FROM WHITLEY STRIEBER

I would like to acknowledge Anne Strieber's very special contribution to this book. She has read every letter that we have ever received, and was responsible for selecting the ones that appear here. The remarkable communication that emerges on a reading of the whole document would not be visible were it not for her rich, open-minded and questioning insight into the visitor experience, and the care and rigorous honesty with which she made her choices.

And I looked, and, behold, a whirlwind came out of the north, a great cloud, and a fire infolding itself, and a brightness was about it, and out of the midst thereof as the color of amber, out of the midst of the fire.

And out of the midst thereof came the likeness of four living creatures . . .

—Ezekiel 1: 4-5

TABLE OF CONTENTS

CHAPTER ONE

The First Communication

This book of letters could conceivably be the first true communication from another world that has ever been recorded. It does not offer a conventional message, but rather something much more arresting: the transmission of its meaning through the lives of the people who had these experiences and the determination to write them down.

Their gift to the world is this: here is the voice of another reality, speaking clearly at last, offering a message and a promise that we can understand and act upon.

And not a moment too soon. Presently, close encounters are the most misunderstood, misrepresented, and scorned of all human experiences, and yet as certainly among the very most important. Over the past ten years, Anne and I have received nearly two hundred thousand letters from close-encounter witnesses. Almost certainly, these letters contain the key to this extraordinarily mysterious experience, and in this collection we have attempted to choose a cross section that would both evolve some rich new questions and provide much-needed answers to old ones.

The collection will put certain shibboleths to rest forever.

First, it has been generally assumed that this experience only happens to a few people. The mere fact that we have received so many letters suggests that the number of people involved must be vast.

Second, there is an assumption that close encounters only happen to people in isolation. If the wife has an encounter, the husband sleeps through it. Actually, most encounters involve more than one witness, and the letters reflect this.

Third, it has been claimed that encounter experiences are only remembered under hypnosis. Nothing could be further from the truth. In the right hands, hypnosis can be a useful tool in helping witnesses to retrieve memories of encounters, but the overwhelming majority of people remember without its aid.

Fourth, it has been assumed that people report these experiences in order to get attention. In fact, we had a great deal of difficulty even finding correspondents who would be willing to allow their letters to be included in this book. People write us to get something off their chests. They do not ask for or want publicity.

It has also been claimed that the experience is an outcome of temporal lobe epilepsy. It has been falsely claimed, for example, in *Parade Magazine*, that I have "admitted" I have this illness. In reality, I have not only claimed but proved the opposite with extensive medical tests, and so reported in my work. We have included one letter in this collection from a temporal lobe epileptic who has *also* had close-encounter experiences in the presence of another witness. Additionally, we have intentionally included the testimony of an alcoholic, not because the close-encounter experience is any more connected with that disease than it is with temporal lobe epilepsy, but rather because it reveals that impairments neither induce nor suppress the experience. It also contains elements of multiple-witness observation.

Some older UFO investigators have tried to promote the notion that all close-encounter experiences are more or less the same, involving people being taken aboard spaceships, examined, and interfered with sexually. In actuality, while sexual contact is commonly reported, this rigid narrative structure is extremely rare except among people who have been exposed to the ministrations of such investigators. We have rarely received letters describing this experience, except in the context of much stranger and more complex interaction.

Until quite recently, the body of letters in possession of the Communion Foundation were the single most powerful argument that the close-encounter experience was both quite real and entirely misunderstood. We did not publish the letters, though, because we felt that they would be unfairly dismissed and derided. That will still happen, but in order to do it now, those who would deny this reality must openly lie. This is because another argument has emerged, this one based on physical evidence, which may now be added to this overwhelming torrent of anecdotes.

For years, close-encounter witnesses have been claiming that their strange visitors have been implanting objects in their bodies. In 1996, these objects began to be located, removed, and placed under laboratory study. The National Institute for Discovery Sciences, using the facilities of New Mexico Tech, determined that some implants removed by Dr. Roger Lier and others consisted of highly magnetized metal, clad in minerals that would have the effect of retarding rejection. Others were discovered by Dr. Lier to be encased in membranes made from the skin of the individuals bearing them. The Communion Foundation has joined together with other foundations to study such objects using the facilities of various research institutes on a par with the New Mexico facility. The first implant they studied proved to be impossible to explain by conventional means. In further support of hypnosis as an effective tool when used correctly, one witness, who had many remembered encounters, was hypnotized and asked if she could remember any implants being put in her body. Under hypnosis, and only then, she described such an event. The area was then X-rayed, the object was located, and it was subsequently removed.

That so much supposedly educated criticism has been directed both against the reality of the close-encounter phenomenon and against the few tools that are effective in helping us learn about it, such as hypnosis, is a grave moral outrage, an assault against the progress of human knowledge by the arrogant and the ignorant.

As if the implant discoveries were not enough, throughout the nineties, hundreds of hours of videotape of UFOs have poured in from around the world, most of it made by amateurs and often in the context of large assemblies of people. This video ended the debate about whether or not unknown objects are moving through our skies: they are.

Taken together, these two developments have changed the situation regarding the letters completely. Denial of the existence of the visitors is now a purely emotional affair. It is absent of any empirical validity. Somebody or something unknown, operating both in the context of high theater and great secrecy, has approached us.

The letters offer us a remarkably pure and efficient means of understanding the message that these very strange others may be bringing us. It is not a message that can be written on a billboard or explained in a few simple words, but it is an astonishingly clear one nevertheless.

As we journey from one bizarre tale to the next, an overall pattern emerges. Again and again, the close-encounter witness has his understanding of the world around him shattered. He is revealed to himself in a new way; the world around him ceases to sustain the familiar meaning by which he had defined it and becomes wrapped in mystery. He is challenged by questions that he cannot bear and cannot answer. Again and again, the plea will be made in these letters: Help me, I'm confused, I cannot understand.

There can be only one reason for such disorientation. At the very deepest level of the mind, so deep it is hard even to remember what happens there, the visitors are inducing stress, and it is severe stress.

Why would they be doing this? Is it an attempt to drive the human unconscious insane? Obviously not. An effort to harm us could be made without expending anything like the resources that are being poured into the visitors' work.

When species are stressed too much, they collapse. They go extinct. Slightly less stress, and they evolve to meet the new conditions. There is no evidence that the visitors are inducing overwhelming stress in the human population. To the contrary, the number of letters that we have received reporting even the slightest physical injury is tiny. We have not yet received a letter reporting that mental illness has emerged from a close encounter.

So, if the experience isn't dangerous, but is stressful, what is the outcome of the stress? Again, the letters make this clear: the stress produces the most profound opening of the mind we have ever witnessed. We don't even know the meaning of the properties of mind and spirit that are evolving in us in response to the presence of the visitors.

It is easy enough to call them psychic powers, but that ignores the fact that they must have some basis in reality or they wouldn't exist. More than the ability to view things remotely, to travel out-of-body, to slip back and forth in time, to levitate, to recall past lives and read the minds of others, close-encounter witnesses begin to see *themselves* in a completely new way.

This experience appears to be, at its core, the outcome of an attempt to wake us up to our own true potential, to enable us to realize that we have been seduced by a lie, and to shake off the chains that this lie has heaped on our shoulders.

The lie is that man is weak and small and stupid, and that the human spirit is guilty or evil or simply inadequate. Scientists, academicians, and the media keep up what amounts to a relentless diatribe against the worth and spirit of man: you have no soul, you have no meaning, your doings are at best venial, at worst unspeakably evil. And history, bloody as it appears on the surface, would seem to reinforce this litany of accusations.

The communication from the beyond that is contained in these letters is quite the opposite. It says that man is more than physical body, that life is rich with meanings as yet undiscovered, and that human beings are rare creatures of extraordinary value.

The hollow, sorrowing people who are the outcome of the barrage of lies that swirl around us are stunned to passivity. They are excellent consumers, because they are by necessity dissatisfied and seeking to fill a hunger that cannot be filled. No matter how exquisite the watch, how speedy the car, how alluring the latest lover, material consumption cannot substitute for self-discovery.

We are a species in poverty, and the visitors are here with a message of hope: you can raise your eyes, and when you do you will see a mirror. This is a magic mirror, for it reflects only the truth. It reveals you as you are: a being so rich with unfulfilled potential that you are literally terrifying to yourself.

Mankind *is* looking up, is seeking to find a way to escape this demonized and dying world. These letters reveal that there is a force calling to us, seeking on our behalf to do what we cannot do ourselves, which is to recognize our own true value, and our own real powers.

Over the years in which I have suffered being the "*Communion* Man," the only things that have kept me going have been the love of my family and the letters I receive. Week in and week out, for the past ten years, we have opened impassioned statement after impassioned statement, and heard there a song, sung in notes of purest gold: We are out here, we are sharing the mystery with you, we will sustain you. As I have faced lying, vicious attacks, one after another, and seen the way ignorance and fear join together to make human beings stupid and cruel, I have had this testament to sustain me.

And so I have joined this great band of sisters and brothers of the unknown, and when I have gone out into the dark to meet the visitors, I have walked with them beside me. Were it not for them, I do not see how I could have continued on my quest. At some point I would have fallen down and not gotten up, were it not for their need for me to rise, and their faith that I would do so.

Deep within the letters, there are certain visual themes that are suggestive of their origin in a truly nonhuman consciousness. Among them are the black sedan phenomenon, in which black sedans inexplicably rush at children; the thug effect, in which readers are invaded by what appear to be 1920s gangsters; the appearance of owls, which often seem to announce the visitors. While riding his bike as a child, one witness was actually hit by one of the mysterious cars and apparently killed. His remarkable story is astonishing evidence of how little we understand about ourselves and the world around us.

The appearance of a large-eyed creature accompanied by two smaller blue ones also occurs more than once in these letters. Why would such a strange detail recur unless people are reporting something really seen?

The whole close-encounter experience has a texture of the improbable that lifts it right out of the ordinary context of "alien contact." If this is contact with aliens, then they are even stranger than the strangest beings we could have imagined. They do not play by the rules of reality but of dream; they are not bound by the careful laws of physics, but by the wild ones of the imagination. Those of us who see them and have the outrageous temerity to report what we see are branded as "alien contactees" by the media and asked dumber-than-dumb questions such as, "What planet do they come from?" and "What are their names?"

They told me that they came from "everywhere," and their names, I suspect, are no more fixed than the wind. As far as why they came here, they said to me, "We saw a glow." At the time I thought that this referred to cities in the night, but now I feel that it was us they saw, our souls like embers, and they knew that we were trying to grow bright.

Then there are those who brand the visitors as evil and dangerous. They would suppress people such as me and letters such as these on behalf of the spectacularly wrong-headed idea that mankind is better left ignorant. But ignorance itself is dangerous, and what we need is understanding so that we can cease to be passive, and so that we can respond.

It is time for those who deny the existence of the unknown to change their position; there is somebody else here, and we do not understand who or what they are. As you read these letters, you will find every assumption you have ever made about the visitors coming into question. You will see them operating from within a context that is totally unexpected and so far unremarked in the literature: the visitors and the souls of our own dead seem somehow bound together. There are some stunning stories of encounters with dead friends and relatives, including one of literally overwhelming significance.

More than one text discusses going out-of-body and finding what is essentially another world, one invested with the technology of the visitors. In the world of the soul, though, even the machineries are constructed of truth. So it is that the ships up close seem like living beings, and the stories that military men tell of crashes and the need to "protect our airspace from the aliens" come to seem like the helpless nattering of blind men attempting to describe by touch what must truly be *seen*. Indeed, the whole folklore of alien encounter as it presently exists seems more like a two-dimensional creature's description of a ball passing through his flat world: he sees only an undulating line. The description is correct, but does not even begin to accurately explain the reality of the ball. Indeed, it can't, because a two-dimensional creature *cannot conceive* of a three-dimensional object.

These letters, however, offer a thunderous and incredibly exciting affirmation of the human spirit and the human mind. Focused though he may be on three dimensions, man is not intrinsically a narrow, three-dimensional creature. Far from it, the letters offer proof that, however imperfectly and haltingly, *we can see*.

Despite the relentless shrieking of scientists, reporters, and intellectuals, steeped in Marxist dialectic and still loyal to its hoary denial of human grace, the truth is that the mind of man can become greater. We can add time to the shapes of the world, and see beyond, and go beyond. The letters call to us; they sing a glorious song: You are more than you think, heed the message of the visitors: look up, rise up, ascend into your own true humanity, find yourself in the stars.

It is entirely clear that the message here is from a higher world: not that we cannot join, but that we can; not one where we aren't wanted, but where we are. And so you will read again and again, amid all the fear and confusion and the desperate passion of the unanswerable question, the haunting call of the visitors.

It is a call, finally, from child to child on a fine young morning: Wake up humanity, come out, join us, and in so doing, join yourselves to the grandeur and simple joy that is all around you.

CHAPTER TWO

CHILDHOOD

My heart leaps up when I behold
A rainbow in the sky:
So it was when my life began;
So it is now I am a man;
So be it when I shall grow old,
Or let me die!
The Child is father of the Man
And I could wish my days to be
Bound each to each by natural piety.

"My Heart Leaps Up"
William Wordsworth

Children and Aliens: Where It All Begins

If you are a close-encounter witness, odds are that your experiences began when you were a child. Often, it seems that the children of witnesses become witnesses themselves, that close encounters run in families.

In general, the earliest memories of encounters coincide with the earliest development of the ability to remember. This suggests, of course, that encounters may happen to the witness even before this, indeed, that they may follow witnesses from the beginning of life to the end.

Childhood encounters are very unlike adult ones, and for a

reason that I believe I can explain from my own experiences. Usually, although not always, the childhood encounter is not deeply terrifying. Children tend to be startled during the initial approach, but they soon settle into a state of wonder that adults must learn to recapture if we are ever to make progress in understanding the experience.

The reason that this is so is that the child's mind is more flexible than the adult's. A child's ego is not developed enough to be threatened by the overwhelming power of the visitors; the devastating fear of annihilation that so characterizes the adult experience appears only as the personality matures. As a result of this, these letters are exceptionally revealing.

Memories from childhood's open innocence are windows, like no other, into the secret heart of the encounter experience. They suggest something very different from adult encounters: there is magic, there is wonder. When one child asks why the visitors are clipping his fingernails and scraping his skin, he is given a wonderful, if somewhat chilling, explanation. Another follows the ghost of his dead brother into the woods, where he is shown a photo album that contains a shocking, deeply revealing secret that concerns us all. In another case, some little boys armed with BB guns blunder into what was apparently an underground alien facility of some sort and have a truly extraordinary adventure there.

What these stories from childhood tell us is the same thing that my own childhood memories have said: to really gain from our encounters, we must balance on the thin line between innocence and wisdom. We must go forth into our encounters with the open eyes of children and the careful minds of truly wise adults.

A Secret About Memory

I have a daughter, age eleven, and this is her account of her experiences, as she told them to me while speaking into a tape recorder.

The first experience she remembers happened when she was ten. She got into bed and had just closed her eyes when a "kind of fog" came over her. Soon she saw the ones she calls "friends" or "comforters." They took her to a stone wall where she felt herself held down, although there were no straps and no one appeared to be touching her. She then felt her "soul" and "mind" go through the stone wall into a gray room. She remembers being on some sort of table, like an ironing board. At this point she wanted to tell me more, but said, "They took a piece of my mind out so I can't remember, because they're not ready to tell you."

She says that when she goes with these visitors, they experiment with her, and "talk to her in her mind." She wants to ask them questions, but somehow can't. They tell her not to be afraid, that they won't hurt her. She feels they really care about her and love her—that they are friends who are protecting her.

My daughter says that she belongs with them and "when she's done here," she'll "go back to their own planet." When she visits the stone wall, her physical form is still in bed, but her "soul and mind" travel to the gray room. She feels as if she's floated up, or carried into a beam of light. In this state, she is no longer wearing her pajamas, but is instead wearing something that feels like leaves.

On these journeys, she feels like she's floating into a ship, but she has never seen one. They take her to a "different planet," where she sees "no other life-forms as we know them." When asked to describe her friends, she says they are little, and holds her hand about three feet from the ground to demonstrate how tall they are. Most of them look wet and gray. Some are greenish, with big black rounded and slanted eyes. Their skin is smooth and "there's a smell like no smell from this planet. They're really thin, except they have little pudgy bellies. They have no body parts, four fingers, feet like ours, but like a duck's, sort of, with webs." She did not notice any clothing or hair on them. "Their planet is damp and we might think it has dirty air, but it's really clean air."

She says when she comes back to her body on the bed, she hears a buzzing sound. She usually thinks it's the clock and tries to turn it off, but it isn't the clock that's making the noise. Then the bed shakes, and she's back in her body. After these experiences, she always knows what's going to happen to her on that day, and things always happen just as she says they will.

Mr. Dinosaur

My experiences with the visitors seem to come in waves, and there are three to four experiences per wave. The "wave" that brought this phenomena to my attention consisted of four episodes, spanning ten months from August 1989 until June 1990. In the first one, I found myself on a dirt road by our house, and was then taken aboard a craft. The second one had to do with paralysis while watching lights coming through windows, and watching these lights move around the house. The third one involved "dreaming" about being on a large disk near a reservoir by our house and being handed a hybrid baby while I was told it was my son. The fourth had to do with being awake one night and smelling "cardboard" while realizing that I was paralyzed from the waist up. I heard shuffling noises next to the bed. I kicked whatever was standing next to the bed, then blacked out.

For over a year, I do not recall any more visitor experiences, but I did have a lot of dreams and paranormal experiences, including levitation.

I then became pregnant, and all was quiet until August of 1992, when one night my daughter woke me up by screaming, "Stop, no!" I leapt out of bed and ran downstairs, expecting to find her in fits, but it was as if she had not uttered a sound. She lay there quietly, sleeping soundly. However, when I turned to see how my son was sleeping, he was sitting up grinning at me, wide awake. I scooped him up and took him into the family room to rock him back to sleep. We must have rocked for twenty minutes, and I was actually becoming drowsy. My eyes were closed. Then an odd thing happened: I got a vision of three "grays" standing in front of the rocking chair. It was as if I could see through my eyelids.

It shocked me, and I opened my eyes to see an intense light coming from the kitchen door. At that exact moment, I felt a flash of searing heat infuse my entire spine. Then the light faded, as did the spinal heat. I didn't even stop rocking. My breath came in gasps and my heart hammered through my chest, but I didn't want to alarm my boy, though I was sure he could tell I was excited. In another few minutes, I decided to get up and put him down, asleep or not. He went down with no fuss, but I was really shaken up, and it took me awhile to get back to sleep.

In October 1992, we moved to the Chicago area. My daughter told me that "Mister" could not find us here, and for the first time in three and a half years she quit jabbering about "Mister and the little Misters."

In April 1993, I woke up to hear what sounded like my daughter running around on the wood floor in the kitchen. Annoyed, I reluctantly got out of bed and looked over the banister down to the kitchen and family room, but saw no one. I peeked into my daughter's room and was surprised to find her sound asleep. I looked in on my son, who had learned to walk a few months before, but he was asleep, too. I got back into bed. Perhaps a minute later I heard it again—someone running around! The thought occurred to me that neighbor kids were running across our deck, but when I got up to look, the sounds stopped. Back in bed again, it began again, but I knew if I got up, I'd find nothing. Anyway, I went down and found nothing again. I wasn't surprised when I returned to bed and heard it again.

The next evening, while bathing my kids, my daughter did something she'd never done before. She told me she'd had a bad dream which involved a "Mister Dinosaur," who walked on two legs like a person. He had two blue trolls with him, and Mister Dinosaur had hurt her arm. I decided not to make a fuss about it, but I checked her arm. It was okay, so I left it at that. I called the Center for UFO Studies to report this episode and waited for the next encounter. I was in for a long wait. Aside from a few restless nights, and a very frightening paranormal experience in the spring of 1995, nothing occurred with the visitors.

On September 12, 1995, I awoke in the middle of the night to footfalls running on the kitchen wood floor again. I knew what I was hearing, and chose to roll over and go back to sleep. Then I had a waking dream in which I was running into the kitchen chasing a child. I chased it into the dining room, where it turned to show its face; it was a hybrid. It seemed to be young. "He" was wearing a hooded robe. I chased him around the table and into the living room, where I caught him by the shoulder and spun him around. I sank to my knees and held his shoulders, and looked into a sweet elfish face. He smiled at me, and I noticed little creases at the corners of his mouth. I asked his name, and then a perfectly ordinary middle-aged woman walked through my front door and, smiling, told me his name. I don't quite remember it, but it was something like "O'dan." Then I noticed several people walking in front of the bay window toward the front door. Then it ended.

The night of October 18, as I was getting ready for bed, I said to myself, "They're coming tonight," feeling absolute conviction. I remember closing the door to the walk-in closet and making a mental note to leave it open a few inches. I woke up later and rolled over on my back, and was immediately hit with the vibrating, heavy paralysis. I knew exactly what was happening and actually formed the words "I know this" with my inner voice. I was looking around the room and beginning to panic, hyperventilate, and my heart was pounding. I was determined to stay conscious, so I closed my eyes to calm myself. I then tried to move, with great difficulty, and to lift my arms in front of my face; I had little control over them, though. I decided to try to wake my husband, and began hitting and punching him, but he didn't budge. I stopped when I noticed light coming from the crack in the walk-in closet door. I closed my eyes again, because I felt panic starting to creep in, but then felt a presence next to my bed. When I opened my eyes I was stunned to find three entities standing there next to my bed, two shorter ones in front of a taller one. I threw my arms up to protect myself—or to hit them—but they simply put their hands up while all three gently brushed my hands away. They seemed to be enjoying this "hand dance." We kept at it for awhile, and I tried to touch them, to feel their skin. Then they began to "sing," making a high-pitched, lilting little noise, with their voices weaving together. It was beautiful! I closed my eyes in utter elation and joy; I knew I was smiling. I opened my eyes to see the tallest one come forward, and I simply reached over and stroked his left forearm with my right hand. It gently held my left hand in its right.

It was very loving and gentle. I managed to say "I love you" out loud. The words came from *my* innermost being and seemed to be louder, as if it was not my voice. The being then began to glide away from me toward the walk-in closet. I didn't notice the others leaving, but they were gone. The being stopped at the door and looked back at me, then tilted its head and smiled! I didn't expect this. I waved and said "bye" out loud. The being seemed so human at that moment, as it lifted an arm to wave back and mouthed the word "bye." Then it turned and went through the closet door. As I watched the light fading, my mind raced, and so did my heart. I was so overjoyed that they had allowed me to see them!

I listened as a train droned down the tracks a few miles away. The light and the sound of the train faded, together with the paralysis. The first thing I did was to check the clock; it was 2:50 A.M. I got up and looked outside, but saw nothing. Then I checked on the kids, and they were both fine. When I got back to my bedroom, I checked my compass, which was okay, so I went back to bed. For over an hour I relived what had just happened, over and over. I was so thrilled and honored that they had allowed me to stay conscious. My eyes burned and teared the whole time.

The beings were pretty typical. The two smaller ones were three to four feet tall and had largish eyes, but not huge or ugly. I was focused from the start on the tallest one, who was nearly five feet tall and had noticeably smaller eyes. It seemed familiar, and the most interesting feature was that it glowed. It was really quite beautiful. The head was very much like a human's, but it was bald.

I am going to back up and relate a very frightening experience I had on April 13, 1995, six months before the experience described above. I was a drunken mess at the time and had quit my job a month before. I was in the middle of an overpowering depression, and when this happened it kind of put the fear of God in me. I can clearly see, from my journal, that it was the catalyst that woke me up to my spirituality.

That night, I could not sleep. At 10:45 P.M., as I lay with my eyes closed, on my side, I thought I felt the cat jump up on the bed in front of my face. I smiled as I felt her fur brush my face. I was going to reach out to her, but when I opened my eyes all I saw was a black void about eighteen inches across, with uneven edges. As soon as I saw this, I realized it wasn't the cat. It began to rotate slowly, brushing my face and right hand with air, and I heard a rushing noise. The quicker it spun, the louder the noise became, until it sounded like a hundred mph wind! I became frightened at this point, and it seemed to jump off the bed on to the floor. Just as I thought it was gone, it leapt back up in front of my face. It seemed furious, and I felt a pulling sensation on my chest. It felt like it was trying to pull my soul out. I started to get mad, and it leapt off the bed again, but about fifteen seconds later, jumped back up and was more intense. This time, it grabbed and pulled on the fourth and fifth fingers of my left hand. This made me angry, and my anger intensified as more pulling continued on my chest. I was now in a rage, concentrating my rage on this thing, directing it like a beam. Suddenly it seemed to detach itself and just disappeared! As soon as it was over, I turned to my husband to see if he'd seen it, but he was snoring softly.

In May, I had entry after entry of spiritual dreams and paranormal occurrences, like seeing things in the house and noises. It was a very active time, and I'm sure it was the turning point. I only have a few entries leading up to the September 12 "dream," and they are all rather profound, when seen in retrospect.

On November 10, I either regained consciousness at the beginning or the end of an encounter, because I found myself in that familiar state of paralysis and simply panicked. I was panting so hard that my mouth and tongue were totally dry. The only thing I witnessed was that same light coming from the crack in the door of the closet. At the height of my panic, I closed my eyes to calm down, and suddenly felt something coming from the bedside along the whole length of my body. It was a gentle, jostling sensation. Then I felt motion, as if I was moving toward the edge of the bed, but I kept my eyes shut so as not to alarm them; I didn't want to scare them. Suddenly the sensation was over, and when I opened my eyes, I was still alone and paralyzed, with the light still coming from the closet. Right away the light and paralysis began to fade, and then it was over.

On December 1, at 1:03 A.M., I came out of a paralysis. When it began, I was lying on my left side with the covers down by my waist. I said to myself, "Oh no—not again!", and closed my eyes to wait. Suddenly I felt something smooth rubbing my open right hand, and when I opened my eyes, I was surprised to find that same hybrid child holding my wrist with both his hands and rubbing the top of his head with my hand! He was looking right at me and grinning, and for some reason this tickled me; he was so cute. There was also a normal little girl with dark brown, longish hair wearing a dress with petticoats. She was about four, and was trying to pry open the fingers on my left hand. I felt the distinct physical sensations of their actions. I closed my eyes for a moment, and when I opened them the girl was gone, but the hybrid was still doing what he'd been doing. I asked him out loud what he was doing, and he telepathically said, "I'm chewing purple gum." I asked him if he'd blow bubbles, and he looked at me with a quizzical expression, as if he didn't understand the question. I suddenly felt very tired and closed my eyes. As soon as I did, I felt something moving under my body from head to toe. I opened my eyes and saw myself being lifted up, and watched with intense interest as the sheets were pulled off of my body and I was drawn out toward the edge of the bed. Then I watched myself being lifted higher and higher, moving toward the bathroom, as if I was being carried by someone holding me over their head. There was a strobelike light, and I was moving toward it. At that point I closed my eyes, and when I opened them I was back in bed and paralyzed, watching the light fade from the closet. As soon as the paralysis wore off, I looked at the clock.

In the early morning hours of January 9, I wrote down all the dreams I'd been having. One was about being in a place with several halls and walls that looked like white tuck-and-roll! I also found myself in a semi-fetal position, pitched forward so my head was down. I could feel gravity, and was fighting to breathe, as I was drawing a dense liquid into my lungs. I got really frantic, until I realized I was somehow getting oxygen, and I was okay. Then I felt very solid, ridged and thick; I felt like a dark red root. I felt the pressure of being encased or impacted into something dense, but my consciousness could travel up and down the length of this "root." I felt safe and somehow very powerful. When I woke up that morning, I was physically exhausted, and my eyes burned for about two hours.

A New You

I will begin by saying that I was ten years old in 1953 and lived in a rural area. My sisters and I all slept in the same room. One night my mother came hurrying into the bedroom to wake us up. She told us to look out the window. We did, and there was a large, hot, glowing globe the sky. It floated any way it wanted to go. My mother was a little scared and told us to be quiet, but as I remember it now, it *was* already very, very quiet outside, like a vacuum. I was watching this globe float around when I suddenly became very drowsy. Here was where my conscious memory ended, until lately, when I started reading *Communion*. The "buried" visit now emerges:

I am eleven years old; I can see "him" floating outside my bedroom window. He is only three to four feet tall. He has a skinny gray body, a large head, large slanted eyes, a line mouth, no nose, ears, or hair. I feel eyes that can "film" everything. I heard in my head, "You have been chosen. We will not harm you." Then there was blackness. Next, I am floating through our living room upright, about six inches off floor. Beings are floating with me. As we float past my mother's room door, I look in her bedroom and see between three to six beings that are glowing from their bodies with a pure white light. They are clustered around my stepdad. He had been crippled since he was five years old. The beings are very interested in his legs. They are holding one of his legs in their hands. I am thinking to myself at seeing *this,* "They can heal Pop's legs!" They did not, of course, but I believed they could. Then there was blackness again.

Next, I am sitting up on a "doctor's table" with my legs hanging over the side. I feel relaxed and curious and excited. "My friend," as I believe this being was, is at my feet, scraping the bottoms of them. That tickled me, so I pulled my foot back, and he laughed with his eyes, and in my mind. They scraped skin off of my arms and feet.

They took a little hair off my head and cut my nails. I asked questions in my mind, but before I could verbalize them, they answered back very softly but directly, "We are making a new you." I asked him, "Are you like angels?" and he replied, "Not as you have been taught." My friend said it was time for me to go, and I would forget their visit. Extreme sadness, longing, and sorrow passed through me, and I cried real tears and begged them "not to forget me" and asked why I had to forget them. His answer was, "Because there are those who will tamper with your mind." Then there was blackness.

I think that added to all this was a flash memory of a big screen and seeing I know not what! It was something about a promise of returning to see me again, and a half-remembered conversation about a "bad time on earth." Then I was floating back into my living room and there, lined up like zombies, was my whole family! I thought to myself, "Why do they all look like that?" Then there was blackness a final time.

Back when I started remembering the visit, I asked my sister if she remembered anything strange about our farmhouse. She acted as if I had opened a long-forgotten dream. I didn't tell her anything about what I was remembering on my own, but her memories made mine even stronger and very, very real. She saw the globe; she saw the face in the bedroom window. I made her describe the face, and it matched the one I saw perfectly. She remembers my mother telling her to get away from the window "or they will see you." Where was I when Mother said that, since we were all in the same room? Was I gone, already taken? I don't know. I do know that these visitors are real, and that they are here. I have strange thoughts about "cocoon"-type human bodies in a "white mist," who will wake up when it's time.

I have felt love, compassion, protection, somewhere from these beings. I see "my friend" with his arms around my shoulders, when I cried about going home. In this there was a pure, joyful and wonderful feeling of being chosen (abducted). There is also a raw, terrible fear that if they do come back, I would simply go insane or have a cardiac arrest; they have probably been in and out of my life like threads on a loom.

I had a dream not long ago about the farm: I dreamed the "government men" came to the farm and made me mad and took my souvenirs that the "alien" gave me. They had men in white suits all over our farm, and the "lady doctors" wanted to give all of us shots; she said it was to "feed us." I politely told her I wasn't hungry.

My mother sent me a letter before she died, about a "vision" she had of "earth's demise." It was total carnage, she said, but we would survive it. In her vision she asked if her children would survive; "whomever" answered, "Yes."

About two years ago I spent some time at my daughter's house. It is in a wooded area. I had a dream while there: In my dream I got up and went into my grandson's room and took his hand, and we went outside in the dark, which I would never, never do. We seemed to float to the front gate, out by the road. We stopped for a moment, and all these "funny-looking children" came out of the woods to meet us. I thought they were so ugly! I "thought" to my grandson, "Don't play with those children, honey; they're too different." Anyway, we were escorted to a saucer-shaped thing at the dead-end circle of the road, and a door was opened in it with a ramp going up. The light inside was intensely bright and "my friend" was there at the door. I then woke up!

Approximately four weeks later my grandson asked me a funny question: "Mamaw, what does an alien look like?" I was flabbergasted; what did he know of aliens? I asked him to draw one on his own, and he did. It's similar to the ones I remembered.

FEAR OF BEING DEVOURED

I'm writing to you because of something you said in *Breakthrough*. On page 178 you mentioned the visitor as having "little spiked teeth." When I read those words my blood ran cold.

First let me tell you briefly a little about myself. I'm forty-two, married to a wonderful lady and have seven children. I was brought up Catholic but from birth have always had an open mind.

I say this because I can recall my childhood with no blank spots. I remember soiling my diaper on one occasion, and not long ago I told my mother that I remembered such events and she chuckled politely, until I described everything about the time to her: the house, the wallpaper, looking down a long staircase as my grandfather read a bedtime story to my older sister. As I was reiterating the events, my mother told me that we indeed lived in such a house, but moved from it when I was twenty-two months old. I'm not trying to impress you with my ability to remember things, because to me it was always the way things were. One time that I must tell you about was when I was four years old. I remember it vividly: I was sitting at the kitchen table in the house we lived in and there were lots of relatives. As I was sitting there I noticed something terribly wrong, for sitting right there on the kitchen stove, in front of God and all my family and relatives, was this creature that I had never seen before. I remember pointing to the creature on the stove, and by now I had everyone's attention. Obviously I was the only one who could see it, which only added to the terror.

My uncle went to the stove and passed his hand right through it in an attempt to show me there was nothing there. I saw his hand disappear as it passed through the being. He turned to me and said, "Look, nothing here." By now I was hysterical and my father took me to my room, got me a cold cloth, and I fell asleep. Everyone thought I must have eaten something that didn't agree with me. In 1958 I think everything was blamed on that. I never again saw that face staring at me, that is, until I was browsing in a bookstore a few years ago and saw the book *Communion,* and right there on the cover was that all too familiar face almost laughing at me. The only difference I remember was that my visitor was wearing a scarf.

In April 1992, my wife and I had gone to bed and had fallen asleep, when suddenly we were both awakened by an intense white light. It only lasted a few seconds and my first reaction was to look out the window to see if I could see what was causing this light. At the same time I realized that it was coming from everywhere, that is, there were no shadows, everything was white, blazing white, and then the light was gone, plunging us into total darkness. Our eyes had to get used to the dark again and the streetlight glowed its dim light into our bedroom and things got back to normal, whatever that is.

It wasn't long after this incident, I would say four months, that I had a rather disturbing dream that left me in a sort of helpless and quizzical state of mind. I was, I felt, aboard a spaceship, in the company of gray aliens that wore long white robes. I was walking with them down a corridor and into a room. In this room I felt I was being introduced to about six young (baby?) aliens; some had long wisps of very fine hair. They were beautiful, but as they gathered around me I had this feeling of dread and suddenly I realized that I was there for them to feed on. And this takes me back to page 173 of *Breakthrough* because, Whitley, these little creatures had tiny sharp, pointed teeth. They began biting into my arms on either side of me; that's when I bolted, I ran for all I was worth. I also had the feeling that they were allowing me to escape, and if they wished I wouldn't be allowed to go. Well, if they were permitting me to leave, I was taking full advantage of their offer.

I remembered finding a series of tubes. Not hesitating, I dove into the first one I came upon, and after what seemed to be a long slide downward, I was on a busy city street, then I awoke in my bed, my wife beside me. I was out of breath and sweating, as if I had actually been running for my life. Unconvinced it was a dream (it was so real and so was my fear), I got out of bed and went to the bathroom to inspect my arms. Nothing, no marks, and I was just as happy. I was very restless the rest of the night. I have never heard of the aliens as having sharp pointed teeth, so when I read it in your book I had to mention my dream. It never recurred.

During this time our younger boys all had bouts with nosebleeds for no apparent reason. Now kids get nosebleeds and we didn't think anything of it until our youngest boy came right out and told us that a man made him have one in his sleep. He said, a scary man. I cleared off the kitchen table and placed upon it pictures of all kinds of monsters, Frankenstein, Dracula, the Mummy, the Wolfman, pictures of E.T., *Star Trek* people, all kinds of pictures of strange creatures, as well as the cover of *Communion*, which by the way I had not placed front and center, but off to the side. My son didn't hesitate to pick out your cover. He was three years old when this happened.

My gut feeling is that this is probably happening to a lot of people, but for a variety of reasons, it is seldom discussed, (not the least of these being one's credibility). I'm not looking for attention nor am I a wacko. I'm not dogmatic over anything. As I did as a child, I still look at the world with an open mind.

PHONE CALLS FROM AN ELF

I bought *Communion* in March 1988 on my way to Los Angeles to sing at my cousin's wedding. I was simply glued to it. There was a strange sense—yet a vague sense of recognition in what you were saying. However, I just couldn't place it at the time. My memories of my young childhood are clear to age four or five, then I have a block from age six to nine or thereabouts. I must tell you that, although I'm uncertain of this, I may have had a brush with the visitors many years ago, however, to my knowledge, not since.

In the wee hours of Christmas morning of either 1958 or 1959, I was six or seven years old. It must've been between two or three A.M., and I was lying in bed in my bedroom in our tract house in the East San Gabriel Valley of southern California, about twenty-five miles east of Los Angeles. I remember I awakened rather suddenly and was quite unable to move; seemingly frozen in my bed. My room was kitty-cornered from the living-room window, giving me a clear view of the Christmas tree and the large picture window and through it the front porch. The porch light was on and the drapes were drawn, giving me a clear view of anything that could have been on the porch. On the porch, seemingly peering into the living room, was a smallish creature I remember thinking to be one of "Santa's elves."

I was naturally excited and wanted to get up to meet or see this elf much closer. I remember being wide awake and desperately wanting to get up, but patently unable to accomplish this seemingly simple task. He seemed to be three or four feet tall and was cream in color. I saw more of a three-quarter profile, and could not make much of any other distinguishing features. I was not that fearful, just excited, and burning with curiosity. My frustration was building with my seeming inability to rise from my bed. I then had a firm sense of, "It's okay now, just go back to sleep." It had a parental tone to it, loving yet directive, however, not either of my parents' voices.

On Christmas morning, as my sister was going nuts, I was

relatively disinterested in all the stuff around the tree. Instead, I was determined to go out onto the front porch and check for evidence of what I knew was there. No such luck. I came back into the house with my parents rather dumbfounded at my spurning Christmas morning presents in favor of running outside to a cool, crisp, California Christmas. I remember telling them that I had had a "dream," even knowing full well that I had been wide awake and definitely not dreaming. My mother mentioned something that I have since forgotten, and I let it go after that, although, I've never forgotten the feeling of the Christmas morning.

For a couple of years after that I had looked for a return of the "elf" on Christmas Eve, but to no avail. Occasionally, I remember being fearful of dark places, waking up cringing when a closet was left open or had clothes hanging in it in a certain way. After some time, whether through maturation, the encroaching megalopolis of Los Angeles, or whatever, I didn't experience any more events.

All I know is that after reading your books, I have found myself compelled to write to you. I have never done anything like this in my life. I felt drawn to the cover of *Communion* like a duck to water. I can think of nothing else but to write you. Here I am at work, supposedly stamping out neurosis and mental illness, and I am a slave to a compulsion that escapes me. I sit down, pick up my pen, and start writing you. Even now as I am entering this into my computer, I feel the identical rush of energy that is exhilarating and at the same time virtually mystifying. I love it. I don't understand it. I don't want it to end.

After I finished the book, that is, moments after I finished the book, I had a sense of longing to have some sort of contact. I had convinced myself that I could face it with strength and courage. However, I quickly realized the denial that I was engaging in when struck by a mixture of awe and fear when my kitchen phone rang; not in the traditional "rrriiinng" but three "tings." This was like a light mallet on a single bell, perfectly even. Well, I don't have to tell you how high the hackles stood up on the back of my neck. All I could hear was a simple voice, "Send Whit the letter."

I was totally fried. I put it to the back of my head and went on like nothing happened, keeping the rough draft of this letter in my left desk drawer in my office, pushing it out of the way and covering it with anything I could find. That is, until two nights ago when the owl activity increased and I heard a solitary barn owl hoot three times, then stop. I thought of the visitors, and then just before turning in, the phone repeated its three tings.

THE BLANKETS THAT LIKED THE NIGHT

I've had two near death experiences, one when I was five and one when I was thirteen. This was as the result of one car crash and a bus crash. Both times, I saw the "tunnel," a light and a field, which I believed was the edge of heaven. The light was very bright and peaceful. Both times, I also heard a "voice" that talked to me through my mind; this terrified me. When I was five, I remember crying to it to let me go home. When I was thirteen, I remember being scared into going back when I was not sure I wanted to. Since then, I've read a lot about NDEs, and I don't understand why I was so frightened by that voice when most people found it comforting.

Last spring, I saw *Communion* in the library for the first time. The face on the cover scared me so much that I went home. I've always been interested in unexplained things, and I couldn't understand why I was so afraid of a book. I had to see it two more times to get the guts to bring it home. Once I did, it was at least a month before I was able to finish it. I also couldn't read straight through *Transformation*. Then things from my childhood that I'd thought were weird began to make sense.

Like many small children, I had a security blanket. It was off-white, and so large that my mom cut it in half. I named one half "Sniff" and the other one "Sniffer." I also remember a bigger, imaginary "Daddy Sniff." I named him this because of the blue coveralls he wore, just like my dad wore when he painted my room. In my mind, all these "blankets" had the same face as on the cover of your book. I remember telling my parents that I'd met Daddy Sniff, and that when the power went out in a storm he'd fix it. I remember him standing on a "ladder" outside that went up to a bright light with electric sparks flying off of it.

I later had another blanket, this one green, that I picked up when the Sniff wore out; I named it "Nicey." I believed that there were many more "blankets" who could fly, and they liked the night. They came from another land and had a different language, some of which I actually wrote down when I learned to write, at three. I also believed that the blankets came to earth because they had to leave their land as the result of a disaster that they were very sad about. I told my parents this and much more, as early as the age of two. The green blanket's face looked like the one on *Transformation*, only lighter, it seemed to me. All the blankets were friendly, and the only one that scared me was the bigger "Daddy Sniff."

I also remember being scared of imaginary buglike things that I called "gerbils." I'd think I heard them under my bed, or in the basement at night. They had large, dark, slanted eyes, and if I wasn't good they'd take me at night and examine me. I never understood this until now, just like I never understood why they said I couldn't tell anyone about them. I have many memories of nightmares about snakes or black bugs with dark slanted eyes being in my room. I believe these are screen memories of contacts I had with the visitors.

Now I believe that the voice in my NDE that terrified me so much was connected in some way to the visitors. The peaceful presence of the light, which seemed to question me before I heard the other voice, I believe was God. I also believe that the visitors know more about heaven than we do.

Aside from my NDEs, I've had an out-of-body experience (OBE) that more than convinced me that there is such a thing as a soul.

I've stopped looking in my closet to be sure that no one is hiding in it before going to bed at night. A sore behind my ear from what I believe to have been a needle of some sort has disappeared. Several times, the visitors have answered my prayers, my fear of them leaving. When I awoke one night to see a red, white, and blue light in the sky, I thought, "It's them." Instead of being frightened, I was filled with awe and adoration. It's that feeling of love and peace that can come from them that we must hold on to in order to overcome the fear.

Since the last time I wrote you, I have had several more experiences which I believe are related to the visitors. The most remarkable one happened last July [1995]. I awoke in the middle of the night, lying flat on my back in bed, in what I recognized as an altered state of consciousness. In each of my hands I was holding a light-blue object shaped somewhat like a rounded bar of soap. I seemed to "know" telepathically that the being standing beside my bed (one of the smaller gray visitors) was teaching me to use these two light-blue objects to raise my energy level until I was almost in another dimension. For what seemed like several minutes I felt my energy fall back down several times, only for me to use my mind to raise it back up again. At this point I mentally began asking the being about the purpose of the exercise, but as I did, the two objects seemed to magically float out of my hands and blend together over the center of my body. The being and the objects then disappeared as I felt my energy slip back down to its normal level.

What I noticed most about this whole incident was that for the first time ever, I experienced absolutely no fear while in the presence of one of the visitors. Earlier last year I had experienced several weeks of sleepless nights due to fear of the visitors, and especially due to the fear of awakening to see one or more of them in my room. At times when I would feel as if they were watching me, I would find myself pleading with them in my head to help me find a way to deal with my almost overwhelming fear of them. I believe that what happened last July was their answer to my pleas. Their willingness to respond to my requests to lessen the fear has convinced me that they are not as cold and uncaring as we sometimes perceive them to be. Though we cannot stop the experiences from happening, I believe that what we can learn by facing up to them will become extremely important.

"I BELIEVED I COULD FLY"

When I was five, back in the 1930s, I was totally convinced that I could levitate at will. In many ways I was a "different" sort of child; for one thing, I was nocturnal, which led to my falling asleep at school a lot. I often told my parents that small doctors visited me at night. My mother told my dad this was normal, as she remembered talking about such visits herself, when she was a child. I explained to my parents that these visitors were the only ones who believed I could fly. I got annoyed and even wept when no one took me seriously.

I was five or six years old when the "sleepwalking" incidents began. I was even taken to a doctor about it. It started when, one dawn, my parents were awakened by the sound of our front doorbell ringing. Dad came out to the foyer and found me ringing the bell while dressed only in the top part of my pajamas, and carrying my bedclothes folded neatly under one arm. Even my live-in nanny had not been aware that I was out of the apartment, and everyone was quite alarmed. As for me, I remembered only that I had suddenly found myself on the front stoop, so I rang the doorbell to get back in.

After a few weeks, the incident was repeated. This time, I appeared wearing only underpants. I had been wearing pajamas over them, which were left carefully folded at the foot of my bed. Once again, the neatly folded sheet was under my arm. The doctor I went to diagnosed somnambulism and said I might suffer from it all my life. I have never "sleepwalked" after the age of seven. However, in those few years it happened every couple of weeks or months. My parents would carefully lock our huge apartment door, but somehow I got out anyway.

We knew every family in the building quite well, so sometimes one of our neighbors who was up early or coming home late would see me and bring me back inside. No harm ever came to me, and nobody ever believed my story, which was that I flew with my friends out to the park on those nights.

Pale at the Window

I have been deeply troubled since childhood by a series of bizarre occurrences that I've never been able to satisfactorily explain away. Nor did it ever occur to me until now that any or all of these incidents might possibly be related. Although reading your books didn't directly solve any of my personal mysteries, for the first time in almost forty years I feel almost at peace with myself, knowing that maybe all of my strange experiences might not be a bunch of unrelated mental aberrations as I always suspected, but *may* be tied together in a way that I never dreamed possible. When I think back over the years, I get the distinct impression that there are countless related memories floating foglike on the extreme periphery of my memory that refuse to come into sharp enough focus for me to grasp, to understand. It disturbs me deeply. I now realize that I've always inwardly felt that I may be, if not involved with, at least exposed to something of an extremely unusual nature.

My earliest, most frightening memory is an ongoing series of incidents that started when I was about four and lasted, intermittently, until I was about eleven. It started when I was living in the Northeast and followed me to the West when I moved there at a young age. At night, after my parents would put me to bed, I'd often see small, very white round faces with huge black eyes staring in at me from outside my bedroom window. Sometimes it was only one, but often it was several. I was quite simply terrified by the sight of these faces, and developed the habit of sleeping with the blankets pulled up over my head so I couldn't see them. I somehow felt protected that way. I slept that way into my teens. These faces in the window followed me when my family finally moved west. I saw them several nights a week almost into my teens. My parents assured me that they were simply a recurrent nightmare, and we didn't discuss it again, but the memory of those faces haunts me to this day. This is because secretly I've always believed that they were really there and that I wasn't sleeping when I saw them. Even now, at thirty-nine, I keep the blankets pulled up tightly around my neck, even in the hottest weather. I know it's the remnant of a habit born of childhood fear.

At about the same time that the white faces appeared in my life, my parents noticed that another mystery was manifesting itself in my room in the dark of night. They'd put me to bed fully clothed in PJs, and upon awakening me in the morning would find that I was totally nude, and my PJs would be in a pile on the floor, very often on the other side of the room from where my bed was located. This also continued almost into my teens.

Shortly after moving west when I was about seven, my family was gathered together in our living room, when suddenly a sort of ashen color washed over my father's face. He looked directly at me and said, "Come on, son." He then led me outside by the hand, directly to the center of the front lawn. He then released my hand and turned his face upward, looking intently into the night sky. I followed his gaze only to see, directly in front of us and about forty-five feet up from the plane of the earth, a dark shadowlike disk, perhaps seventy-five to one hundred feet in diameter and about 1500 feet removed from us. It clearly was rotating slowly as it moved from our right to left, quite noiselessly. I knew it was rotating, because on this disk's surface was a row of round illuminated portholes/windows, and they were moving from right to left. After this disk passed out of sight to our left my father said something like, "Son, we've just seen a spaceship from another planet." At this time we proceeded back into the house. The incident, to the best of my recollection, was never discussed again.

In 1970, while still living at home with my mother, I was awakened in the middle of the night by what felt like several pairs of hands sliding between me and the mattress. I was lying on my back with the hands going under me from my right side. My head was turned to the left so that I couldn't see who/what was touching me. When I attempted to look right, I was terrified to find that I couldn't move my head; I could move only my eyes. Apparently my entire body was immobilized, because my attempts to escape whatever was touching me were without result. I was staring at the curtains and the wall to my left, completely horror-stricken. Even in that state I found myself attempting to analyze my situation and myself. To my surprise I could scream. I couldn't form words with my lips, but I surely could scream, and I did so frantically.

Whatever was in the room with me made several attempts to lift me off of the bed, but it seems that my incessant screaming might have forced a change in their plans, for they plopped me back down on the bed and began to withdraw from beneath me rather quickly. A short time later I discovered that I could again move my body freely. I immediately looked to my right, but the cause of my terror wasn't there. Instead I saw my mother hurrying into my room with arms outstretched, coming to comfort her screaming son. Through tears and heavy breathing, I explained to her what had happened. She assured me, as usual, that it had merely been a nightmare. I knew at that moment, and I've known all these years, that my experience was quite real. I was wide awake and struggling to reason out what was happening to me at the time. It was not a disjointed dream. I was being kidnapped or abducted, and I knew it. Now I'm wondering if anything happened during this attempted abduction that I'm not remembering.

In 1972, while standing outside of my home with some friends engaging in small talk, I saw six points of light moving together as if in formation. They were moving from my left to right at a fairly high rate of speed. When I pointed them out to my friends, we all agreed that we should get into the car and follow them. We drove down country roads for several miles, keeping the lights in sight. At one point, one of the lights broke away from the main group and curved off to the right. We decided to abandon the original group and follow the loner. Eventually it appeared to descend and disappear behind a hill ahead of us. We drove until we reached the hill, and drove around to the side that the light had disappeared behind. What we found was a small valley about the size of a football field, socked in with a dense fog. From somewhere within this foggy area was an incredibly bright light that illuminated the entire little valley. I stopped the car and we all got out. I feel as though we were just about to discuss walking down into this fog to find the source of the light when, quite inexplicably, we all just got back into the car and drove home without another word. We never discussed this incident again and, to the best of my knowledge, none of us ever attempted to return to the location of the fog-shrouded light.

On the morning of May 2, 1991, as I was lying awake in bed at about 6:25 A.M., I heard a loud clicking noise that drew my attention toward the bedroom door. As I stared out into the hallway, I saw two bright flashes that momentarily illuminated the hallway. I got up, dressed, and went through the entire house. I found nothing that could have caused these flashes. I then went outside, expecting it to be storming, complete with lightning flashes, but the sky was cloudless.

On Monday night between 11:00 P.M. and midnight, I woke up into one of the strangest experiences of my life. I was awake, could feel and could smell and think and reason, but I could not see. There was frantic activity going on over me and to my right. I mean frantic with a capital "F." I could feel and sense the rapid movement of one or more somethings all around/over me. It was so rapid that this rapidity scared the hell out of me! I was being "swarmed" over and touched all over. What scared me most was that, through it all, I was struggling to reason out what was happening to me. I experimented with every part of my body to see if I could move; I couldn't. There was a flashlight a few inches from my head, but I couldn't make my arm respond to my mental commands. Nothing worked, so I tried calling out to my wife. All that came from my throat were gurgling sounds, Then quite by accident, I discovered that I could manage a very small, snakelike squirming motion. After a period of time, I managed to squirm until I was touching my wife. I tried bumping her with my buttocks and it worked!

I did this two or three more times with no apparent results; she wasn't going to wake up. Then I went back to reasoning. During this entire time the activity hadn't ceased for a second, and I was on the verge of losing myself to panic. I began to mentally compose a list of all the things that could explain what was happening. I considered the possibility that maybe I was dealing with visitors à la *Communion*, or perhaps it was something I'd eaten that evening, The thought crossed my mind that I was the subject of demonic possession. At that exact moment, all the activity around me abruptly halted and my vision returned, and I was wide awake breathing heavily, but feeling like a ton had been lifted from me. I was drenched in sweat. I raised my head to look at the clock and it was exactly 12:30 A.M. I was on my wife's side of the bed up against her so tightly that any movement on her part should have resulted in her falling out of bed. To my astonishment, she said the next morning that she felt she'd had the most restful sleep she'd had in a long time!

Something is definitely happening in my life and I don't know what. It seems to be gaining in intensity since I've read your books. I'm not complaining. I have a deep feeling that this is all something that I've known was coming for a long, long time. In a bizarre sort of way, I want it. I go to bed every night now scared to death, but still wanting something to happen. I want to face whatever has entered my life and learn not to be afraid of it.

CHILDREN DISCOVER AN UNDERGROUND FACILITY

Thanks for the interest in the underground UFO base near Hamilton Field, California.

When we moved there the area was still fairly rural. In front of my home was a large hill that soon became the playground of the children in the neighborhood. We spent thousands of hours playing on it with our BB guns, which closely resembled real rifles; that is a point to keep in mind. There were four of us. Always, at the top of the hill at a fence line, we would be stopped by men dressed in plain khaki pants and shirts, who would tell us we could go no further. Playing Indian one day, we did cross the line, and found an underground base of vast proportions. We literally slid under the fence on our stomachs, going from bush to bush and tree to tree. We slipped into the forbidden area. In a small gully we encountered a shimmering mirage effect that reflected the sides of the gully as mirrors reflect the sides of a magician's box. Going through the mirage, we found a large tunnel that led deep into the hill. Thus, we entered the base undetected. There was no road to the tunnel, only a well-traveled path that connected the tunnel to a gray-green, ratty old house set back among the trees. This house also served as another entrance into the base. The tunnel led into a landing area divided into four bays, equipped with a conventional bench at the back and a variety of metal parts hanging on the walls. We examined the benches, and then entered a large door that was fully four inches thick, approximately twelve feet high, and seven feet wide. This size is the most important thing to recall.

Entering the base proper, we were fully functional and free to go where we wanted, under no restrictions. A groove ran down the middle of the tunnel floor. This divided into four other grooves, one of which led to what seemed to be a landing bay. The floors, walls, and ceiling were all the same gray color, unlike the rest of the base. There were benches across the back of the bays that were constructed of conventional wood, but I didn't recognize any of the tools or odd bits of metal that were lying on the benches. Large pieces of metal, which I assume were spare parts, hung on the wall, very similar to a conventional mechanic's garage.

In Bay #3 was a conventional UFO, which looked like a round, dome-topped craft with square, angular vents with a tube angled in the back and a sooty black inside and tripod landing legs. Keep in mind that the bays were quite large, and that I have mentioned some details of our exploration out of sequence. Bay #3 also had occupants, but more of this later.

The green house: After leaving the bays, we spent time in a living quarters area, then entered a large area split between incubators on the left with small clusters of alien babies, who were covered with short brown fur. The green house on the right contained wide-leafed tropical plants in pots arranged in rows. On the leaves of some of these plants were what appeared to be brown beetles, about 2 feet 2 inches long, with no legs or feelers. In this area, I was attacked by three smaller "bugs," as I have nicknamed them. We fought a short, vicious fight, and I escaped back to my friends, then we started to leave. They were playing with some of the babies at that time.

While leaving, one of us shouted that he saw monsters, and we ran back to the bays. In the bays heading to the tunnel exit, I saw my friend standing like a post in front of Bay 3. Calling to him to come on, he said he couldn't move. Thinking he was just scared, I went to get him. He alone of us didn't carry a BB gun. Looking into Bay 3, which was crowded by the UFO and dimly lit, I saw movement, and hid behind my rigid friend for cover. Then a Mexican stand-off occurred, when the beings in the bay were afraid of the BB gun, which I fired to no avail. I was afraid to move out into the open. Finally, a boxlike device moved out from behind the UFO, and my vision blurred and I got groggy.

Finally, I noticed a weird, full-sized figure coming toward me like a robot with a red target painted on its chest. I found out later it was body armor. The target was designed to attract fire to the center of the armor, the safest spot. At this point I got really groggy.

I came out of it surrounded by a circle of larger beings,

heavier built, all staring at me every which way I turned. I panicked, and broke out of this ring to the tunnel where my other friends were. We tried to mount a counterattack to get to our friend, but panicked and ran outside. At a loss about how to get our friend, we hollered for him to come out several times. He finally did, but in a badly shaken state. We calmed him down, and then started to climb the hill toward home. As we climbed, I saw a large, bright silver object just above our heads, and the heat became intense, so intense that I began to stagger. Finally, I realized we couldn't make the top, and I told my friends to hide in the brush while I tried to make it home for help. Behind, I could see the beings coming after us. I continued on a short distance, and then fell. I crawled into the brush to hide. Behind, I could see the beings standing over my friends. Vaguely, I was also aware of them standing over me, telling me, via telepathy, that I must "forget everything, forget, forget . . ."

I awoke at sunset, with my friends kicking me awake and, slowly, we went home exhausted and totally unaware of what had happened.

The first of many times I was on this base, I saw a room like an airplane hangar, full of what I feel was a row of machines. The outside of these machines were made of six feet by six feet panels, fitted together. They were great, featureless lumps in a vast hall that was noticeably warmer than other areas. I also saw a conventionally furnished room, where aliens and humans were freely mixing with each other. They had human males working with them as "front men," so to speak. I and other children went through a selection process, and those selected were trained to build equipment for the bugs on an assembly line. I was good at this delicate work. Finally, it came down to four boys working as a team with one boy from our group among them. Also, I recall that a girl I knew was on the base. She was raped, and it meant nothing to the bugs. I believe they used kids to do hand-sizing for the tools used. We were the same size as the adult "bugs." The bugs as a group were not a society I'd like to see in control of earth. They are vain, humorless, and indifferent to human pain, mental or physical. They operate with a totally different set of values. Their society is regimented, and I doubt if any crime exists. Each has his job, or place in life, and they are ordered; individual initiative is lacking. They function as a team, working toward a goal.

However, they learn fast in one-to-one fighting. Still, it is copying whatever happens, such as throwing dirt in their faces to confuse them. I am getting ahead of myself here, and maybe only clouding my points, but having worked with them I also got to know something of them. Finally, reduced down to four boys working as a team, we met on the base under the tutelage of a human male named "John." John was always interested in expanding their power in human society. This, as a child, angered me. I always held part of myself free from them. Inside I knew it was wrong, that they were undermining the country and my society.

How does one cover a lifetime of periodic contacts in one short letter? Or explain all that I've seen? There were uniformed Air Force officers at times, as well as a giant, long-furred biped creature used for labor. The bugs were five feet tall, gray, slant-eyed bipeds with large heads, stout short necks, gray uniforms, and some had a red circle on their chest. This is rather a rambling discourse I must admit, but it's written as it occurs to me.

I have returned to the site of the tunnel, and it has been covered over. The old house is torn down and replaced by a new, modern house. The once-barren hill is covered in part by new houses now, except for the old entrance area, which is still wild. A cellular phone unit is nearby, which has been taken over by the U.S. government.

My sister, also, has now begun to remember the bugs, the base, and the UFOs. Her children, it also appears, are now involved with the bugs, but as yet refuse to discuss it freely. The social issues of my old friends are complex, and we are bitter enemies now. This all started with territorial disputes over the base area. They started telling me I couldn't go up on the hill anymore, to which I replied, "Want to bet?" We literally fought wars over the issue of territory on the hill. Over about eleven years, my memories have slowly returned on this issue, and when I am comfortable with one thing, something else comes up.

How big is the base? About two and a half miles by one-half mile, on two levels. It is a huge base area with warehouses, machine shops, barracks, labs, assembly areas, etc. I have continued to remember more rooms and machines. I estimate about fifty bugs were there when I was there. I played with the younger ones, and for a period of time, visited the base daily during summer vacations. Some friends found a patch of ground that was sodded over. Why would someone sod over a small patch of a large, bare hill, except to cover up something? No doubt I have missed several vital points, but this should give a rough overview of things. Never, ever get the idea that bugs are kindly, space-age gods. They are not. Accept them for what they are, no more, no less. Do not ever trust them in word or deed.

P.S.: Let me describe a "light" I saw in the living quarters. It was a glass globe, three feet in diameter, hanging down from a foot-long, six-inch-thick glass column out of the ceiling. In it was a whirling mass of silver sparks, like fireworks, constantly moving and changing shape. It was fantastically beautiful to watch.

Silhouettes from Childhood

I am a fifty-three years young, wife, mother, grandmother, and psychic. I enjoy good health and like to think I'm reasonably sane. I was taken several times as a young child, mostly at night. I'd awaken to an inner voice calling my name. I'd walk out the door and down the hall, and they'd be waiting at the bottom of the stairs. I can still see their silhouettes; there were always two of them. Those visitations stopped when I was about eleven. As I went about my life I tucked the memories away deep in my mind, chalking them up to "childhood dreams."

Six years ago, I had an incredible experience, another visitation. "She" told me that I'd be given information and gifts and would be allowed to remember. My encounters are both physical and spiritual, and my life has been profoundly changed. I believe that I have some sort of receiving device implanted in my left ear because of the ringing, buzzing, and clicking sounds in that ear. It's also the side of my head from which I receive channeled information. I had my neck X-rayed a couple of years ago, and was startled to find that I have a fused vertebrae in my neck. Also, my eyes are unusually sensitive to light.

Now about the gifts! They told me I could paint, even though I had no interest in the art world. I thought they were jesting, since they do jest. However, I bought a couple of brushes and some basic oil colors. I discovered that not only could I paint, but I was good at it. I sold my first painting for $150, unframed. I've gone on to have four, one-woman shows, and I sell my paintings about as fast as I can paint them. Then they told me I could write poetry. I tried, and now have about twenty works published. I don't believe that they gave me these talents; I believe they brought my latent talents into my conscious mind. The amazing thing about all this is that I'd never tried any of these things before. The topper is that I'm working with only an eighth-grade education. I find that just about everything I try comes easily. Also, my psychic ability has increased.

The beings are very spiritual and worship the universe in its entirety. They believe all that exists is of God and is God. They believe that all material beings have a soul and body and that the body has its consciousness just as the soul does. The body and soul are a marriage of understanding and mutual benefit. They don't look at death as we do. They say nothing dies, it simply becomes something else, and they believe that even the universe is in a constant state of becoming. They don't understand why our culture separates itself from God, or thinks of God as external, instead of getting in touch with the God that we really are.

My last physical encounter with them was about seven months ago, when I was taken to an underground installation. I was told that the group was doing genetic research relating to bone abnormalities, including bone cancer. I was informed that about half of all medical breakthroughs in the past twenty years have come as a result of their research. This research is channeled into the consciousness of those who will use it to benefit mankind.

They disclosed that in the year 2029 the earth will tip on its axis. This will be caused by a natural alignment of the planets, during which the combination of their gravitational pull, along with that of the sun, will pull earth from its orbit. The visitors are doing the measurements and calculations that will be needed, so that they may assist at that time if necessary. This is another reason to educate the public about the presence of our visitors from other worlds.

CHAPTER THREE

Abduction

Soon, full soon,
Dost thou withdraw; then the wolf rages wide,
And the lion glares through the dun forest:
The fleeces of the flocks are covered with
Thy sacred dew: protect them with thine influence.

"To the Evening Star"
William Blake

Alien Abduction as It Really Is

Early UFO investigators, taking their cue from hypnosis-induced recall that they have structured around a desire to present a believable and understandable image of the encounter experience, popularized a very specific narrative of alien abduction: the victim is taken from his home, placed on an examining table in a spaceship, then subjected to various medical maneuvers, including the use of huge needles and bizarre sexual experimentation.

My own initial remembered encounter in December 1985 was an abduction, and I was intruded upon physically to such an extent that I had painful injuries afterward. But there was something about my encounter that did not fit the image of emotionless, robotlike aliens engaged in some sort of scientific study. To the contrary, my experience was terrifically emotional. My visitors were tattered and nervous and rough. They were scared of me. They were upset by my screaming. Hardly robotic, their actions seemed to be as much an effort to force me to look at them and remember them as it was any sort of scientific study.

I was not too surprised, therefore, when letters started coming in, to find that the so-called typical abduction scenario is rarely reported.

Much more often, what is reported is far more bizarre. One of the abduction stories recorded here, for example, involves a sort of medical experimentation—but it's much stranger than anything ever reported thus far. Another ends with the visitors putting in an appearance as Prohibition-era gangsters, complete with blazing tommy guns! In two other cases, the Blessed Virgin is involved. In one, she appears as an alien. These two letters are included because they characterize a whole Marian subculture that exists within the phenomenon, and one of which I am a part: At one point, I was very much ready to believe that the strange being depicted on the cover of my book *Communion* was the prototypical mother of us all.

Stories like these are much more typical of the real situation. There is nothing neat, clean, or even very comprehensible about these stories. If they are tales of actual alien abductions, then they suggest that we understand literally nothing about what they might be doing to us.

What is clear is that abduction has a powerful psychological and spiritual effect; indeed, during the experience, the line between body and soul often becomes so blurred that it is impossible to tell the level on which the events are unfolding.

The myth that witnesses must be hypnotized in order to remember experiences such as these is again put to the test. Like the overwhelming majority of my correspondents, none of these witnesses has ever been hypnotized, and as you will see, the reports are, as always, sprinkled with multiple witnesses.

Given that these few letters represent thousands of others, the great majority of which are from perfectly normal people who are reporting experiences that are literally beyond the bizarre, it is incredible that no part of the federal health system, neither the Centers for Disease Control nor the National Institute of Mental Health, have made any effort at all to understand this phenomenon. Normally, if even a few hundred people reported symptoms as unusual as these, there would be a concerted effort to discover the cause. Not in this case. The few studies of abductees that exist are very limited in scope, and the government's interest in the welfare of these citizens is nil.

Is this simple negligence, or do they not study the problem because they already know its cause?

The first two letters included here are typical of another subgroup, which we believe are reports of the initiating phase of abduction. In these cases, the witness is conscious of the approach of the visitors, but then there is a blank. The first of the two, in fact, contains the most vivid description of the visitors in all of our records. I have spoken personally with its author, who is a grandmother and a schoolteacher, and is as ordinary and down-to-earth as she can be. Apparently, something happened to prevent her from being placed in an unconscious state as the visitors approached her. The fact that the incident unfolded in the middle of the day suggests that the visitors are much freer to operate in our world than we have realized. Indeed, their imperious conduct in this case might indicate that they place themselves as a higher authority, with rights over us similar to those we claim over the lower orders of earthly creation.

My correspondent, in this case, ended up with hours of missing time. She ends with a haunting question: "Where did that time go?" Nothing, not hypnosis, not natural memory, not her dreams, ever suggested the least hint of an answer to that question.

It does not mean that there is no answer. The rest of the letters represented here explore what might be hidden in that darkness, and tell a remarkable tale indeed.

CAPTURED BY THE VISITORS

In 1976 I was vacuuming my living-room floor at about noon. Suddenly I felt quite ill and thought I was going to vomit, so I sat down on the couch to see if the sick feeling would subside. I then saw that I was not alone; there were three strange little people standing alongside the couch, just looking at me. I froze with fear, as I had never seen anything like them before, not even in the movies.

Two of them were short and fat, about four to four and a half feet tall, with broad faces and enormous black eyes, but with only a hint of where a nose or mouth might have been, almost like a pencil drawing. They had wispy bits of brown hair at the back of their heads, and they didn't have blue suits on like the ones you described in *Communion*; instead, they were wearing brown shrouds. These, I knew instinctively, were the workers. The other was female, thin and about five feet tall. She wore a black shroud and had black wispy hair at the back of her head. Her face was very elongated, with huge, dark, piercing eyes, and once again just a hint of where a nose or mouth would have been.

The tall thin one started to speak to me with her mind, and told me I was to go with them. I answered with my own mind that I wouldn't go. Somehow, telepathic communication seemed perfectly normal at the time, and I felt quite comfortable communicating that way. This doesn't mean that I wasn't frightened—I was beside myself with fear. She kept saying, "You must come with us," and I kept refusing. She then said I could go free, and I got up off the couch and crawled along my hallway to the front door. When I got there, they pulled me back with their minds until I was on the couch again. They let me go again, and the result was identical, except that this time my husband was standing at the front door. I clung to him, and I will always remember how the sweater he was wearing smelled.

They pulled me out of his arms and back to the couch and once again told me it was useless to fight, as I had to go with them. The two workers seemed to be busy doing something all the time this was going on, but I have no idea what it was. The next thing I was aware of was the sound of my husband's car pulling up to the house. I heard him come through the front door and down the hallway, and at this point, I noticed that the visitors were gone.

When my husband walked through the door, I didn't believe it was him at first; I thought it was another trick. It took me about fifteen minutes to really believe that they were gone and that my husband was home. My next shock came when I found out it was 5:30 P.M. It seemed like it should be 1:30 P.M at the most. I wonder, where did that time go?

An Abduction Begins . . .

Sometime in the fall of 1973, I believe that I had just turned thirteen years old. A total of six teenagers in two rooms would make for many interesting stories on its own, but nothing like the story I am about to tell. I shared my room with my sisters. We would chat amongst ourselves until Mom and Dad would become enraged at the noise of giggling. Threats of impending punishment were a nightly ritual. We would will ourselves into slumber to avoid the temptation to speak.

It seemed that I had been asleep for hours when I heard a noise that sounded like jingle bells. *Jing, jing, jing,* I kept hearing it, but it was not a noise coming from a dream that I was currently involved in. I wasn't aware that I had been dreaming at all when I noticed the noises generating from somewhere outside my window. I had listened to the jingling for a few minutes before deciding to look out the window to check it out.

As I think back, I am intrigued by my completely calm response to what I was looking at. Directly in the middle of the yard where the swimming pool used to be was a large metal saucerlike vehicle. It was almost as if the occupants had thought that the cleared terrain had been prepared for them to utilize as a landing pad. The vehicle was much smaller than what I would have guessed.

I noticed a hatchway appear from where there was previously a seamless place. It just appeared out of nowhere like an invisible orifice. There were four or five small kids who walked out of the ship onto the lawn. They were much smaller than I was and were no threat to me. I would estimate that they were probably the size of a five- or six-year-old. They were wearing body-hugging dull silver/gray suits that seemed to cover them completely from head to toe.

I noticed two of them take something from nowhere. I quickly saw that it was a ladder of some sort. I perceived it to be a rope ladder with metal hooks on the ends. I became amused at the simplicity of the tools in which a seemingly advanced group of beings were forced to use to enter my domain. I heard the metal ends of the ladder as they gripped tightly on the windowsill directly in front of my face. I saw them with my own two eyes right in front of my nose. My heart started racing as fear began to consume me. It was really happening, I not only saw it, I heard it as well.

I saw the fingers of the first to reach my window, as they reached up to pull themselves up by the ledge. These were not people fingers. There were only four of them and they were a different color. They seemed to have a bulbous look to them. It was when the head started to peep up over the ledge directly, two inches in front of my face, that I lost it. The head was donning no helmet. It was completely hairless with wrinkles, like frown marks across the brow line. The color was that of a dead person, kind of ashen. It was about to pull itself up to where we would be eye to eye when I became so terrified that I was no longer able to witness this scenario any longer. I flew out of the room screaming. I couldn't understand why no one was waking up.

Ed. note: The correspondent offered no further memories.

A Place of Upheavals

When I walked past the bookstore, it was only out of the corner of my eye that I saw the cover of your book staring out into the mall. I stopped and looked at it, not believing what I was seeing. My first thought was that it was a work of science fiction, and that the poor author would never know just how close the depiction of the being on his cover was to the real thing.

The similarity between encounters you described, as well as a being you described, and my own came as a shock to me. I had never heard of anyone else experiencing an encounter while sleeping, and am both relieved and excited to find that after fifteen years I'm not alone after all.

My episode occurred in the early 1970s. One night while soundly asleep and in a dream state, the dream was suddenly interrupted by a loud noise and the appearance of a stark white face and head, which faded into and out of focus several times, directly in front of me. Although I felt I was fully conscious, my eyes were closed. I remember struggling desperately to escape from the face, but I couldn't, nor could I open my eyes until, abruptly, it released me and I awakened in a state of profound terror. Several hours passed before I dared try to sleep again, and this only happened after I convinced myself that I'd experienced some unusual form of nightmare. Finally asleep, I remember dreaming again; all was normal. Then, again, a loud noise interrupted the dream and the white face appeared uncomfortably close to my face. Again, I was able to see it though my eyes were closed. This time, the face didn't fade in and out of focus as it had before, but appeared clearly and with great strength. I was unable to turn away from it.

The upper cranium was enormous, as were the black eyes, and its slender face tapered to a long pointed V. It looked as though it had been carved from alabaster. Its skin was not supple and hinting of underlying flesh as ours is, but was thin and stretched tightly over its bony structure. Its nose was long and pointed, and its mouth, which was slightly open, was nothing more than a straight slit. It didn't have lips. Its first words to me were, "Don't be afraid. We will not harm you." There was something strange about the voice, but I can't remember now exactly what it was. There was also something especially strange about that mouth, which I remember quite well: it didn't move when he spoke.

My next recollection is that I was somehow in their craft with them. There were only two of them and both were dressed in a type of long silvery tunic. My impression was that they were not small, as were most of the people you described, but were large. I felt humbled, or perhaps childlike, in their presence. The lighting was subdued in this large circular room, but I was able to see consoles flashing with subtle lights. They brought me to one of the consoles, and a holographic image appeared. It was a grid of moving lines. (I didn't realize it had been a hologram until recent years.)

"There are positive and negative forces in the universe," one of them said, "and these forces flow freely next to one another."

The grid then showed a depression in one area, and the lines appeared to lock together in a tangle. It said, "There are spots, however, where the interchange is not smooth and the forces become locked together. In these spots you'll observe an abundance of upheavals: fires, earthquakes, murders, illnesses, and catastrophes." The picture of the grid ended, and the speaker said, "You have been brought here because you will know whenever you're in the vicinity of such a warp. You are to leave that area immediately. Do not attempt to convince others of the reality of your sensing; you will not be believed. Simply leave the area." They didn't give me the impression that they'd chosen me to receive this information as the result of any particular greatness on my part; in fact, they seemed rather bored with me. My impression was that they were merely doing their work, and that I was one of a number of others who would or had already received patient instructions to do something almost as simple as coming in out of the rain.

Next they led me to what looked like a podium and opened a book. There were a number of names in it, and following the names were birth dates and, in some cases, dates of death. The speaker explained that these were the names of others, some of whom had remained in such warning areas. I was able to read the handwritten list and the names appeared to be perfectly common, American-type names. I attempted to memorize what I was seeing as a sort of "proof" to take with me when I left, and looked closely at one of the names and its corresponding dates. They realized immediately at which name I was looking and said the man had lived in Chicago over a small toy store which had once been a pawn shop. As they talked, I could clearly see the storefront, and believe I could recognize it today if I located it. They also offered additional information about the man's family, including the name of his brother.

Next, they showed me visual images of a couple of example areas to avoid, giving me the impression that they weren't sure I was understanding what they were telling me. One area was a field of strawberries, another a large flower nursery. I have since located the nursery, but have never seen the strawberry field. Then, abruptly, they were finished with me, and the white face released me.

I am unable to recall if they warned me not to speak to anyone of the episode, but I've felt a prick of guilt, not to mention foolishness, on the few occasions when I've tried to tell others my "dream."

The following morning I was quite distressed and described the "nightmare" to a close friend. He felt I was overreacting, and thought the entire event humorous until I told him the name, birth date, and date of death that I'd memorized from the book. The name and birth date belonged to a past friend of his who did live in Chicago and also had a brother with the name I supplied. My friend immediately tried to contact this man by writing, but after several days the letter was returned marked, UNABLE TO DELIVER. To my knowledge, my friend was never able to locate the man. It's been a number of years since I've been in touch with my friend.

I have never seen the great white beings again, although I've begged them, in my mind, to return. To hell with the terror, I miss them.

The second thing I want to tell you is that I've just

completed writing a book. One of the main characters is based upon the white being, another is a Native American. I wrote this story feeling that my mind had woven a good and fabricated yarn. Recently, though, I had the great honor of receiving as a guest in my house an Apache medicine man. Through him, my husband and I have discovered to our shock that portions of my book, including words I thought I had invented, have basis not only in ancient rituals but also in the Mayan language, which our friend speaks.

I feel compelled to tell you that there are portions of my book which deal with owls, a cornfield, a communion called "Cura Elow," and a triangle. I had no knowledge of the importance of any of these situations when I wrote the book, most of all of the triangle. I became aware of their significance and possible meaning only after talking with my Apache friend, and reading your book. Let me suggest that the presence of a triangle may not be so much a question or riddle in itself, as it is the indication that an answer has already been given.

One final word: Since the early 1960s I've wondered about the origin of a scar on the palm of my right hand which simply appeared one morning. I think I shall now stop wondering. It is a triangle.

THE BLESSED MOTHER HERSELF?

I'm not sure of the exact date, but I believe it was mid-fall 1973 when I had my first visitor experience. I'm sure of the time of day; it was 8:00 P.M. At that time of my life I was living a very hectic schedule. I was a young mother with two small children under four years of age, and going to college part time at night. I didn't like going out that night to do my grocery shopping, but that was the only night it could be done. I do remember that it was a cold, dark night with few stars visible in the sky. The trees were bare without their leaves. When I reached the stop sign at the end of my tract, I noticed some strange lights up in the sky approaching me, not terribly high up or really low either. What was so strange was that they were pulsating, not flashing, varied in size, and making a trianglelike shape in the sky. There were three lights, red, white, and green, and I believe that the green was the largest, though I couldn't be certain. I definitely remember them being different sizes. They seemed to approach me quite swiftly, and then suddenly they stopped. I became frightened. I looked to see if anybody else was out to see this, or if any car stopped to look at it. I believe only one or two cars passed in front of me. This lasted for maybe two minutes. I don't believe I lost any time. During that time I remember freezing in fear. These lights seemed to be looking right at me and the area all around me. Suddenly the triangular lights turned swiftly to the right and sped away with incredible speed. No airplane or weather balloon could have moved that fast. I just froze and had to talk myself into getting the shopping done. I rationalized that I must be seeing things. Somehow I completed my shopping, but I had this awful, cold fear the whole time I was in the store. I just wanted to get home.

When I got home I told my husband what I'd seen. He showed some surprise and concern, but not much. I know he was tired, and maybe I was seeing things. We went to bed early that night. Despite my apprehension, I definitely remember not having any difficulty falling asleep that night, but I had the most amazing dream:

What I first remember is floating down my upstairs hall in a sitting position with my legs straight out in front of me. There were these little black people, about five or six of them, alongside me. They were all smiles. Their eyes were round and they had no hair. I don't remember if they had clothes on. They seemed to be all black and shiny. I felt anxious, but not very fearful. I remember feeling stiff and only moving my eyes from side to side. I thought that I was frozen with fear, but it's strange, I wasn't terrified. When we came down the stairs I could see that there were others who were swiftly roaming around the downstairs part of my house. We have a sliding glass door off our kitchen that leads into our backyard. I don't remember if the door was open, or if we went through the window. They were taking me up a ramp, I think, into a large round spaceship. Before going on this, I saw two round spaceships on either side of the one I was entering. In 1973 we didn't have the arborvitae trees bordering either side of our property. The craft extended over into our neighbor's property. I think I should add that we've never had any problems with our backyard lawn, nor have my neighbors.

What I remember next is waking up and feeling pressure around my nose. When I opened my eyes I saw this huge white bug with enormous eyes. All I could think of was a praying mantis. I was on my back on some table looking up at this thing. I remember becoming very frightened. I don't know if I spoke or not, but those eyes looked very angry to me. In my mind, I believe, flashed "space creature." Next, I remember pleading verbally or with thoughts, I'm not sure. I wanted—no, I demanded to know what was going on. I had to understand. I knew I was begging. Those eyes turned away from me. The visitor turned to a table to my left, then faced me again. Those eyes seemed to go right through me. What I remember next is quite interesting:

I wasn't on the table anymore. I believe I was walking, not floating. I was approaching someone amazing. I also forgot to mention that I felt that the huge white bug was female; why, I don't know! To continue: This new person or visitor was luminescent white and floated. She was showing me an underground cave. She was very gentle and kind, and was trying to soothe my fears. This may sound strange, but I was "trying to meet her halfway," so to speak. All I remember thinking was that I had to know and needed to understand. She started to walk me deeper into this cave. We passed the little black people, who were all smiles while carrying boxes, I think, from one area to the other. I had so many questions, philosophical ones, I think. She was wonderful. I can't remember exactly what we discussed.

When I woke up from this "dream" I felt wonderful. It was still the middle of the night. I don't know what time it was. All I know is that I felt so loved and cherished, and that everything was harmonious and good. I forgot to mention that when I met this visitor I wondered if the huge white mantis was playing a trick on me. It crossed my mind that maybe I was seeing the Blessed Mother herself; I just don't know, but I put my fears aside.

THEY CAME AT NIGHT

At a fair I was at, there was a man selling books. Eight or nine tables were stacked with them and nothing interested me except one about Nostradamus by Erika Cheetham, then a "sensing" suggested a move further down the tables. There, my eyes made contact with your *Communion*. Going through it, nothing really impressed me, until at page 254 the pages seemed to be stuck, so I started to read the part about the triangle. It almost put me on the floor! The table supported me, and the hair on my jaws was very stiff, a trait from my father that comes when we're very emotional.

On my upper left arm is the mark of an inverted triangle with a "T" in it. Years ago a "sensing" suggested that this would be tattooed on me: a green isosceles triangle around a red "T." On page 128, you really hit me again. Why, of all places, are we both marked on the left arm? Suffice it to say that your books *Communion* and *Transformation* have helped me understand more about my past sixty years than anything ever read by me.

My first enlightenment came in 1948, during a journey to San Antonio to enlist, after moving here from Iowa. I went into a very small bookstore close to the bus station and bought the *Dialogues of Plato*, containing the "Apology," "Crito," "Phaedo," and "Symposium," and got wrapped up in reading it and missed my bus. You took this a step further and really explained it on page 174 of *Transformation*. Anyway, let's go to page 121, where you describe the "others," the names I call the beings with me.

Late in 1934, pneumonia got me. Early in 1935 it doubled after coming back, and the doctor thought my time was up in this dimension. They had me downstairs in a west room on a daybed, close to the west windows, which allowed me to look out on the backyard. This was my favorite place to play, especially the two big cherry trees. The rest was maybe an hallucination from the fever, but three separate times? One evening at dark, two beings came to the windows in an elongated envelope. They and their conveyance were like Northern Lights, and they asked me to go with them, and each took a hand.

When they touched me it was like being lightly tickled, and very pleasurable. I couldn't see them as being like us. The odd part was that you could see through them, but they were there and we talked without sound.

They took me to a huge dance pavilion where pairs were waltzing. My impression was that the males were dressed in black with white trim. The females were dressed the opposite way. There was no ceiling, but crystals hung above and made music beyond comparison that sounded to me like wind chimes. Years later, hearing my first wind chimes, the memory returned. They seated me on some filmy substance, but I remember none of the return trips to my home. On the third and last trip, one of the females came to me and asked why I was there. My reply was that they brought me. She said for me to return, as they had a purpose for me. She asked why the dark frightened me, and told me that both the light and dark were my benefactors, but they would make them equal. She reached up, took a crystal and placed it in the palm of my left hand. For a second, it resembled the "Star of David," two triangles, then melted and ran between my fingers.

My mother took care of me days, and a retired RN did at night. One morning she told my parents that just before dawn, when she went to move me (they turned me two or three times an hour); she found that the bloody cough had stopped, the fever was gone, and though every other window was frosted with the intense cold, the west ones were clear, with the moonlight coming in as bright as day.

In a few weeks it became apparent that my night vision had changed; my eyes would adjust to allow me sight when it was pitch black. The doctor said it was a phenomenon of the fever. The odd part now is that our daughter used to plunder her toy box in total darkness, and also the dresser, to find some item she wanted.

My second gathering, if the first was true, occurred in the summer of 1941. Our Boy Scout troop had hiked three miles east of town and camped on the banks of Crooked Creek. On the first day, three of us built our lean-tos together, and the second day we all worked to build a dam in the creek, which wore us out so sleeping was no problem. In the wee hours of the third day, a strong urge to urinate awakened me. (Unlike yourself, crowds and cities overwhelm me. The outdoors and nature, with her creatures, is the place for me.) Anyway, instead of merely going to the creek bank, a very bright shaft of moonlight enticed me to walk due east to a glade in the woods. The shafts of light began a strange movement, and there was my envelope, and something like an oval cylinder; it just seemed correct to go into it. In an instant there was the pavilion, but the "others" were not dancing. They seemed to be conversing, but not with me, when a lady asked me to "step on the dais" because I was their "conductor." For some reason this seemed plausible, so I complied. Above me was a triangle that gave off lights beyond description. It descended, and traversed the entire length of my body until it went out of sight below my feet. It stopped for a second at my midsection, and surged into my back.

Next morning, I and all my bedding were outside the lean-to. The most unusual thing was that two large moles in the middle of my back were very sore, and stayed that way for a couple of days. These moles had appeared six to eight months after my birth, and my parents thought of having them removed. At about the age of one and a half, my mother went with a friend as a lark to visit an old woman who claimed to be a seer. They'd left me in the car with another child and a friend. The seer was talking to another woman, then paused and told my mother that "her baby had two contact points on his back and they were under no circumstances to be disturbed." They seemed to get smaller with my getting older, and I sense they may disappear shortly before my third, last, final meeting with the "others."

My belief in my Creator is absolute, but I don't accept the physical church. Each time we sleep it is death, as there is no control by us, yet upon awakening one sees the sun with the perpetual energy of our Creator. What was the purpose of my being the "conductor"? My soul is as common as the earth. Why the dreams once a decade about levitation? My father often called me a "throwback." As an explanation, he said my ways were ancient. He considered sugar or honey to be great medicines. He always treated my deep wounds with only sugar. I asked him one time why he did this, and he answered that they told him to use it; who were they? This dummy "throwback" should have asked more questions!

I read a book about a man by the name of Edgar Cayce who could go out of his body and travel. He claimed to have seen a tremendous oak that held all earth's truth since the beginning of time. I wish they'd let me read it. [*Note:* The "book" referred to here, of course, is the *Akashic Records*, which Mr. Cayce claimed was in actual physical existence, buried in a line with the sunrises at certain days, in a secret labyrinth close to the Pyramid of Giza.]

The Caddo Indians of Texas told the first Spaniards of the "blue lady" that visited them to tell them about our Creator, and told them her name. She, in reality, was supposed to have never left Spain and belonged to an order that wore blue; however, it's a matter of record that she often spent days in a sort of cataleptic trance.

Some friends have recently told me that they feel intimidated by me at times, not physically, but by something coming "through" me. Others have said that they "feel safe" in my presence; unlikely with one so basic. All this is maybe mendacious, but one item is absolute fact: your books unlocked some of my doors.

GANGSTERS FROM THE BEYOND

Last night I finished reading *Transformation.* Similar kinds of phenomena found their way into my life over the years. In September 1987, I shared the speaker's panel with you in Brewster, NY, and wanted to talk with you then, but the opportunity slipped by. Like yourself, as you noted in the paperback edition's latter pages, I've no great enthusiasm for the standard UFO cults and their prosaic mentality about the extraordinary. I've lectured at various groups and appeared on a couple of TV shows, but still I find it impossible to share some of the more astonishing aspects of "personal encounters" with visitors.

One of the things I was told some years ago is that "encounters are replicas of something else." In other words, many things people say they see and experience are effects; they are not necessarily the living source from which the effects originate.

From this and other very dynamic kinds of experiences (yes, some being bizarre) I came to a startling realization that many apparently intensely real contacts with visitors and their craft are indeed effects of a crossover into a living and universal spiritual dimension, no less real than anything imaginable in the material world and experience, but rather infinitely more so. Because of the phenomenon of the human mind itself, I was given a lengthy demonstration about the nature of emotional and mental perception. The things I've seen and experienced constitute such a vast tapestry that I almost wish to liken it to that: a correct side, which reveals our past and future in a single, timeless correlate.

I wish to share one particular experience. One night in 1972 I was undergoing a departure from my physical body. As I lay in midair suspended, slightly to one side, over my bed, I observed my room much as you described your similar occurrence. Into the room walked two little people, one of whom carried a black satchel. The satchel was opened and formed three sections, its inside portion appearing like red velvet. It looked just like a silverware display case. One of the little fellows removed a metal probe and came over to my body; the other took a boomerang-shaped translucent object and moved it in a semicircular motion around my head. My scalp was simply laid open and moved over my head, and the other person took tweezers and began implanting what appeared like rubies and sapphires into various portions of my exposed brain. When this was done, the scalp was simply replaced and the boomerang-shaped object again moved over the "wound" and the job, whatever it was, was done. These two little fellows walked to the doorway and just a moment faced each other, bowed and departed, as if having said, "Mission accomplished."

On another occasion, these same two little people were dressed like 1920s thugs, and came into the same bedroom with old-fashioned tommy guns, aiming at me and blazing away. I felt the pellets bounce off my torso, and for several days I had pains in the chest. It was a trial, one could suppose, because this occurred prior to the first experience related. They seem to delve into nascent fears, test a person, and then return with all kinds of amazing compensations.

I was finally taken to the source where many of my visitants exist. I'll never know until my book is published, if any publishers are that courageous, how many other people have undergone similar experiences. Having been a former theological major in a seminary in my twenties, I became an émigré from traditional religious teaching, but when I underwent many astonishing encounters with both angelic and not so angelic types, I realized that orthodox religious concepts were but a mutated façade for dynamic underlying realities, and of course my world ended not in a bang but a whimper. But the pursuit goes on.

The Black Box

I have been keeping this secret from the age of eighteen; I'm twenty-seven now.

One night in my bed at my parents' house, I was contemplating falling asleep and the time was a little after 10:00 P.M. As usual, I was thinking about life.

Then I was pulled up and through the ceiling head first. I reappeared in a large room with a high ceiling, or no ceiling. I seemed to be floating. I looked up and saw long ropes hanging down, lots of them. They were thick, maybe six inches around, and that part of the room was full of them. They didn't reach the floor, but hung down. They didn't taper at the ends, but were chopped off. I was close to one rope, and it looked like it had a white cocoon webbing all over it. As I moved toward it, I got the idea that it was dirty and I didn't want it to touch me. Then I was moving away from the ropes.

I approached a box. This box was about three feet high and five and a half feet long, and was flat gray in color. As I came to it, I was on its left. I remember looking down at it and seeing no seams. I was maybe ten inches from it. As I looked at it, I got the idea of a "Pandora's box," and the thought of opening it came to mind. Since it seemed like a dream I thought, "Hey, why not?"

Then I was somewhere else in a lighter room, and I must say that it was round. I couldn't see well, and was trying to focus. It was hard to see. I was just standing there. I felt other people were around me, but there was no sound or color.

I saw someone across from me, about twenty feet away. He was white and misty and I knew it was a male. I remember thinking, "He's just like me!" Someone was observing me close up on my left, out of my field of vision. As I became aware of this I said, "May I look at you?" The answer was, "No, you may not!" The tone of the voice was female and also firm, as if talking to a child. I obeyed as if I was indeed a child. I was thinking, "Wait a minute, who *is* that to tell me what to do?" Then I realized that it wasn't a dream.

I started to turn to my left to see who it was that had spoken to me. As I did, I started to panic and screamed bloody murder. I saw the top of a bald white head, and holes for eyes. I felt myself start to rush away.

Then I was back in bed, feeling a great disappointment and feeling a little pissed off. What the hell did the box mean? Then I began to remember the rest of it, and denial set in. I wrestled with this for awhile. I came to this conclusion: How could I, or anyone else, conceive of what had just happened to me? Who could I tell? No one. What should I do? Nothing. That's exactly what I did all this time, until now.

Liquid Blackness

It's been two years since I first had a desire to write to you. I'm not sure why today's the day I'm finally doing it; maybe it's because I've just finished *Majestic* for the second time, and find it as captivating as the first time. I'm thirty-two and have been happily married for eight years to a wonderful guy. I have two beautiful kids, ages seven and four, and am lucky to live on a 113-acre hobby farm outside of the small town of Norwood, Ontario.

I've had an experience with a being that has changed my life forever. When I was nineteen, I had my first OBE. I was living alone at the time and it totally blew my mind. It was so utterly incredible, yet so obviously real. I spent the next ten years alternately fighting them off and enjoying them immensely, depending on my frame of mind. The fear of the unknown stayed with me, but fortunately I had brave periods also that let me explore the astral world to a certain degree. I learned through my own experiences that death is but a new beginning, and earth as I knew it was a reality so small compared with the vast realm of the beyond.

I should say here that, to my knowledge, all my hundreds of OBEs throughout the years have been conscious ones, meaning that they've all occurred in the state just before sleep, where I am fully conscious and aware of the paralysis, the vibrations that occur, and of the actual separation. Personally, I feel that separating from the physical is the most exciting and incredible part of all.

On the night of March 15, 1989, I went to bed and fell asleep normally. Sometime during the night I awakened to find myself softly bumping against the ceiling, already separated from the physical. I knew that I was "out," and I was aware of my surroundings. I couldn't believe it. I felt as if I was being lightly tossed about and had no control over my movements. I then felt myself fall to the floor on the other side of our bed, and I lay there looking up at the bumper pad of our water bed. Just as suddenly, I felt myself falling (being pulled?) down through the floor to the basement below. I was upright, and facing the stone wall behind our wood stove, moving slowly sideways as though I was exploring the stone. I recall being very confused as to what in the heck I was doing.

It was then that I felt the presence behind me, and I had a feeling of nervousness so intense that I felt my heart would explode. As I said to myself that I wouldn't turn around, I felt myself being turned around. I "saw" a being standing in the middle of the open room, approximately fifteen feet away. A telepathic voice asked if I was afraid. I thought back that I was very nervous, and "he" said not to be, everything was fine. I then felt myself floating to his side, and I found myself looking at the most incredible sight I had ever seen. The being was most definitely not human, but I don't want to label it alien either; I don't know what it was.

I'm 5 feet 3 inches, and I was floating in the air. He (I knew it was male) was at eye level with me. The basement was dark, but I saw the grayish skin, the overly long skinny arms, the small hands with the long fingers. I took in the large, funny-shaped head and the little slit of a mouth. I registered the fact that it was wearing a uniform of some kind. Yet I'm surprised that I noted any of these things, because I can never forget the look of his eyes: huge almond-shaped pools of liquid blackness. They were so large that they seemed to wrap around the sides of its head. I saw an incredible intelligence within them that was so overpowering I remember no more of its physical appearance. The last memory I have of that night is the comforting feeling of kindness and love and staring into those eyes.

I have no memory of returning to the physical, which is another first for me. I recall waking up in the morning terribly tired, and it took great effort to get up and get my daughter off to school. I was outside waving to the school bus and aimlessly wandering the front of our property when the memory of the night before returned. I couldn't believe it and I couldn't stand it; it was impossible that it had happened to me.

I've had OBEs over the years that would curl your hair. I've experienced fear that most people could never comprehend. As the still-fresh memory of that being coming into my house and pulling me out of my physical body to meet him, with a familiarity that I can't explain, came to me, it was too much for me. I found myself searching the lawn and the driveway for burn marks, spaceship tracks. Actually, I don't know what I was doing.

I told my husband about my experience that night. He had a physical sighting of a UFO in summer of 1978, so he understood to a degree, but as with OBEs, the experience is your own and you're the only one who can accept it and live with it and try to understand.

Approximately thirty-six hours later I developed a really strange eye infection. My eyes were terribly bloodshot and drippy, with a thick yellow discharge. It cleared up about five days later. I recall being overtired for weeks afterwards.

One other thing was odd. The first week in May my husband had a business meeting and I didn't want to stay in the house alone. I had to force myself to go to the exact spot in the basement where it happened. One brave day I actually meditated on that spot to try to remember what had happened after I saw the being. I still can't remember. When I went with my husband to the meeting, the hotel where we stayed had an indoor pool, so while my husband was at work I donned my bathing suit and called my brother and his friend to meet me there. I was in the pool with my kids when they arrived. The first thing they did was freak out over this mark on my back. It was on my left shoulder blade on my back, approximately three inches wide and five inches high, and looked like I'd been burned. There were funny crisscrosses through it, and from the appearance of it, it looked as though it should have hurt terribly. I felt no pain. I had no idea how long it had been there. That "burn mark" didn't fully disappear until Christmas, eight months after it was first noticed.

It took a long time for me to accept that a being not of this earth knew me and had come for me in the night, and it was all done astrally, as though he'd materialized and called for my soul.

I've become more of a spiritual person since that fateful night, far more interested in my inner self and the state of mankind than in the psychic in general. I only hope and pray that people will realize that earth is just the school of life, and that it's true that love and helping each other is all that matters. I believe that incredible intelligence is awaiting us all, and I am trying to be worthy of their interest in me.

CHAPTER FOUR

ENCOUNTERS

Judge not the Lord by feeble sense,
But trust him for his grace;
Behind a frowning providence,
He hides a smiling face.

"Light Shining Out of Darkness"
William Cowper

The Classic Close Encounter, or, Who's That Coming Out of the Wall?

If the witness phenomenon could be said to have a central motif, it is the close physical encounter with the visitors. Close encounters of this kind may or may not also involve abduction, but the axis of the memory is the initial approach itself. Sometimes the witness has no feeling that there has been an abduction at all.

Among this group of letters are some of the most amazing we have ever received. One, for example, involves an entire family having an ongoing encounter over a period of days. Others suggest the presence of lost worlds, while others seem to indicate some kind of symbiosis between us and the visitors that is totally unremarked in the UFO literature, which seems closer to medieval tales of interaction with incubi and succubi.

Two of the other letters mention the black sedan phenomenon, which, like the owl, wolf, and "threes" phenomena—in which things appear in groups of three, or happen at 3:33, or take place three nights running—form a small part of the otherworldly grammar of the experience, which often seems like the remains of a lost theater of meaning, or even a symbolic language.

The close-encounter letters, in general, serve as a powerful reminder that the visitor experience is profoundly connected to physical reality—although, probably not in the same way that *we* are. In any case, is not an imaginary experience, nor some sort of insubstantial hallucinatory myth. In the end, the experience is *real*.

But in what way? Two of the letters contain descriptions of a strikingly similar being with a shield on its chest, which I also saw at one point, and which I associated somehow with Mars. Am I describing fact, shared dream, some sort of unconscious archetype or cultural conditioning? Presently, it isn't really possible to tell, but the tone of the letters speaks on behalf of verisimilitude and truthfulness.

Some of the most powerful of them involve attempts to literally drag the soul from the body. This experience has been infrequently described in the literature of occult attack and alien abduction, but it occurs often enough for us to receive twenty or thirty letters about it a year. Those included here are typical of the genre.

As real as these experiences are, superficially there would not seem to be much evidence that they have a conventional relationship to what we call fact. Or do they? All of the bizarre side effects of contact, with visitors walking through walls, appearing and disappearing at will, levitating, and taking us on incredible hallucinatory journeys, can actually be explained more easily by the application of technology than they can in any other way. And yet, there is the persistent impression that this involves movement between levels of some sort, that to enter the world of the visitors is to enter a level where physical reality is not as fixed and absolute as it is in our own.

Is this an actual place, where spaceships can traverse vast

distances in an instant and visitors can walk out of our dreams into our living rooms, or are we looking at side effects of a technology that invades the mind and overwhelms its powers of discernment so completely that it, quite simply, becomes detached?

These letters begin with one that suggests the visitors are within us, journey through many of the incredible varieties of close encounters, and end with three that comprise a surprisingly large number of the letters I receive: encounters with other human beings under extremely unusual circumstances. One of these letters involves being drawn out of body to an air force installation, and two involve encounters, under very unusual circumstances, with me.

"It Is Me Within Thee"

I've only told one other person, my husband, about the uneasiness and fear I've encountered time and time again since the age seven or eight or possibly earlier; I can't be certain. I'm frightened of putting my thoughts about all of this on paper. I've always felt "monitored" in some way. The words "tagged" and "tracked" have been a vivid memory ever since I can remember. Unless someone has walked in my shoes, they can't possibly even come close to being able to understand such events.

You mentioned a smell like cheese. I've also noted this with one exception: it's more like a "yeasty" odor like rising bread dough, not overtly offensive, but certainly not pleasing. You also mentioned the "voice" coming out as a "deep bass sound."

To me, it was in spurts, like a computerized, synthesized version of language. Their mouths remained closed, and their language came from somewhere within, or on, them. It was monotone, nonexpressive and unnatural sounding to me.

One very eerie detail you wrote about was the appearance of a wolf, and his howling. I got a gut-wrenching feeling in the pit of my stomach about that. For years on a nightly basis, I had recurring nightmares of a huge, glowing red-eyed wolf, the size of the one in "Little Red Riding Hood." These "dreams" were always the same, to start with, and also I knew I hadn't been asleep when they occurred. I was deathly afraid of going to sleep at night, and often I'd find myself sitting wide-eyed in bed with the bedclothes disrupted, soaking wet from sweat, and with my heart leaping through my nightgown. I was too scared or too numb to scream, no matter how hard I tried.

Until the past eight years or so, I've always sensed something coming up from behind me. I'd turn quickly and gasp, as if someone had brushed my shoulder or blown air onto my neck. I can also remember many episodic occurrences, as currently as three years ago. I'd be in bed and very much aware of my surroundings: how the pillows were arranged; my robe lying in a certain way on the chair; our dog asleep on the edge of the bed. (No, he didn't react any more than your animals did, which led me to believe I must have been hallucinating, since animals would certainly have been aware if they observed someone!) I'd feel uneasy, as if I needed to turn over on my other side, as if somehow that would make it go away, this same panicky feeling that I'd sensed as a child with the elf visits. I'd then hear our clock radio playing on the bedside table and logically know that no one had turned it on; my husband had left for work and it didn't just go on by itself. I'd hear music, interrupted by a mechanical type of voice, but I could never remember what it said. I'd feel the mattress depress and the springs pressing down, and I knew someone had sat down beside me. But I was paralyzed, and couldn't utter a sound, swallow, or move in any way. My eyes were open, though I was always facing in the opposite direction from where "the presence" remained. It would be there about ten minutes, it seemed. I'd pray, from my Catholic upbringing, that God would spare me this and protect me. Like you, I also often prayed to Mary. I'd slightly relax, and then boom! I'd somehow jerk hard, and be back in control. No radio would be on, the dog would be sound asleep, and it was utterly quiet. It would be much later; I lost two to three hours each time. I'd have a headache and dizziness and would feel "flu like." I felt hazy, and as if I was moving in slow motion. When I did things, I'd stare blankly a lot until I'd become aware of it and tell myself to stop.

I also went through disturbing episodes of nasal bleeding. In my case, they became projectile and frightening. I associated sleep with nosebleeds.

As a very young child in the 1950s, I can remember a cartoon on *Lunchtime Little Theatre,* a TV show. It was called, "Love in an 'Airship.'" I'd become rigid with fear, and would run out of the room when I saw it, not knowing why. There were three characters: a woman, a man, and one that horrified me whom I referred to as the "Big Beard Man." They went up into the sky in a helium balloon and I remember nothing else about it. I'd like the chance to view this again to see if I'd react in the same way.

You mentioned being inquisitive and asking many questions

in school. I shared that particular trait and would often drive my teachers crazy in a verbal checkers sort of exchange/match. At around the age of six, I developed an intense curiosity about the medical field and desired to be an anesthesiologist. My parents and grandparents could never understand why I'd choose that specialty. No one had undergone any surgery. I certainly hadn't even been given Novocain at that time and no one we'd met were doctors, except our family doctor. I remember being torn between that and the idea of becoming a nun. I was devoted to the religious part of my life, and attended weekly Catechism classes, made my communion and confirmation, etc. I remained devoted until my second year of high school, at age fifteen. Then my interest gradually waned. I felt sadness about that, but I'd lost my faith with the nuns and priests as examples. I had an endlessly questioning mind, filled to the brim with what they considered to be bizarre questions, and I thirsted for answers. I'd try to pry answers out of them, as if they kept secret and hidden truths from us. I was usually told to "believe and all would be well." I couldn't accept that, and I stopped wearing my scapular that year. When my daughter was born, I didn't have her baptized or take her to church, because I wanted her to be able to choose for herself. I have, however, remained close to God in my head and my heart. But I don't share the universal beliefs about this or about creation, or about what I consider to be "symbolic" stories from the Bible. I have yet to learn of a religion that truly shares my own beliefs.

I felt "different" from other children, and had difficulty in forming relationships with them. I remember always drawing as a kind of a tension reliever; even as an adult, I occasionally do so. The pictures I seemed to focus on were of a group of people: One woman carried a small child in her arms. The child was crying and pointing at an object in the sky, a huge, illuminated roundish thing that at that time I viewed as a "planet with ringed circles around it." Later I grew to believe it was a large craft.

I've awakened with unexplained bruises, cuts, specks of blood, etc., many times. Recently I had a V-shaped skin removal from the inner nailbeds of my fingers, on my right hand. This has occurred many times. My husband is fully aware of this. They hurt and are deep, and heal slowly. I also have a specific set of symptoms at the same time: lethargic and "flu-like," feverish chills; mild nausea; swollen eyes and body pain; bruises and odd rashlike areas that itch intensely, all over. This lasts several days to a week or so. I get that odor/scent, which vividly surrounds me. It's that yeasty, burnt, methane-like gas stench, and it makes me ill. It's almost mold-like, in a way.

I've also had several EEGs done. One, done about eleven years ago, showed "increased uptake in the left temporal lobe, indicative of seizure." At other times, totally normal readings were given. Your "temporal lobe invasion" theory makes perfect sense to me.

I feel this may cause a short-term physical abnormality to be present, as on my EEG reading. How can all these exact similarities be utter coincidence?

I have episodes of having had an abundance of "electricity" in my body. I've had hair dryers, clocks, watches, TVs, etc. malfunction or break during these times. My husband can detect these episodes, at times, when I'm unaware of them. There's a "humming" feeling, a "buzzing" that stems from within my body, and is detectable as sort of a force field around me. If he puts his hand two to three inches away from my skin, the hairs on my body stand on end. I sense the humming, and he feels extreme static discharge. We've mentioned this to many doctors and have never been able to get an answer.

Some events include my grandmother, brother, and husband. I'm thirty-nine, and look back on these events with the knowledge that they dealt with something outside of our world, possibly paranormal phenomenon, or hallucinatory experiences or mental abnormalities. But if that's true, why do seemingly normal people report such detailed and similar experiences? How many of us have dismissed these patterns as "daydreaming" or "imagination"? There is no doubt in my mind that I've experienced visits, as you described in your book. That was the reason my husband bought it for me, because I felt disturbed and isolated about this and needed to know that someone else had experienced them too. What is happening?

During one of these episodes, I recalled the words, "It is the Me within Thee." Could that mean that we are a part of them, that they are a part of us? Are we one and the same? I'm quite sincere, and I believe within my heart and soul the validity of my feelings.

HINTS OF LOST WORLDS

I too have had some very strange experiences in my life. They began in June of 1957. Although I can't remember any visitors or actual close encounters, I'd like to share some of my experiences and dreams, as I call them. I didn't realize what I was going through, but the last strange thing that happened to me started about five years ago and lasted for about three years. I, like you, would wake up at night and see the walls of my bedroom and hallway on fire, or what I called "fire with no flame." The walls would be filled with what appeared to be a bright white or yellowish light. My first instinct was to yell "Fire!" to my wife. At other times, I'd just lie down around 12:30 A.M., close my eyes, and then it seemed that in a couple of seconds I'd hear a loud crash or bang in my head, followed immediately by a bright spark of white light behind my eyes. I'd wake up and my heart would be pounding as if I was frightened. I'd also see two white lights, one slightly higher than the other, flying or floating across my room in a descending motion toward the floor. These lights would go right into the floor and disappear. I'd jump out of bed and feel the floor, but it would be cool. The lights were about the size of a silver dollar, very bright white.

I would have what I called a "dream," although I felt that I was totally awake because I could move my eyes. My body would be completely paralyzed. I couldn't yell or scream, but wanted to. I could feel the pressure of something or someone coming toward me, then I'd feel pressure on top of me, and then I wouldn't be able to see. I never saw anyone, but when I'd come out of it I could feel my body tingling all over, and I'd finally be able to scream, and would wake up my wife.

One night I had a "dream," or what I called a very real dream, about two young girls. They appeared to be female, with dark hair and very pale white skin and large, slanted, catlike eyes. One had blue and the other green eyes. They had small facial features, one-piece bodysuits fitted very tightly, and we were in a world with a beautiful spectrum of colored plant life. These colors were colors that I've never seen in our world. This place had two suns, one very close and one very far away. It was the most beautiful place I'd ever seen. Every living plant seemed to have its own glowing light within it. These two girls took me into this perfect structure, like a cottage, into which they entered and told me to follow. I could hear their voices, but their mouths didn't move. I followed them through a doorway opening that I had to crawl through, because it was too small for me to walk through. We got inside. It was a very dull and dark room and something scared me, I don't know what, but I asked them to let me out. They just laughed in a high-pitched tone and left me. I never saw them leave through any opening. I then woke up in a very numb state. My body was tingling and my heart was pounding.

I now live in the outskirts of a small city in New York, far away from any country setting, and this has all stopped for now. I still smell burning paper or cardboard at night sometimes, but I never can find the source or origin of this odor. In every place I've ever lived in my entire life, I've smelled this burning paper smell, with a touch of something sweet mixed in.

One day about five years ago I felt compelled to draw a picture of magnets in a circular pattern, with a rotating magnet in the center. My original drawing had just one north and south pole rotating counterclockwise, but for some reason I think it needs two. I don't know what it's supposed to be, except that at the time I told my wife that this was how a flying disk is propelled. I don't understand any of this.

As I read *Communion*, these things started coming back to me. It was as if I'd told you about all this before, it was so similar. In the sixties, on a hot summer day in June or July, my brother and I were lying on our back lawn in the country watching a gold-colored ball hovering high in the sky for hours. My mother asked us to come in for lunch, but we said we wanted to watch the ball some more. She said it was only a silly weather balloon and to get in and eat. We did, and when we finished our lunch I ran back outside to watch the gold ball. The ball was moving in a zigzag pattern, kind of like a hummingbird moves, toward the woods, and then it disappeared above the trees. I told my grandfather about the ball and he showed me some pictures in a book about flying disks that he was reading. My mother told me not to listen to him, because he'd been weird ever since his stroke. He told me later in life, on the day before his death, about a continent that had joined north and south America with Europe.

This was called Mu, and its neighbor was Atlantis. He said that these people destroyed their continent with a nuclear device and went to the stars to live. They started on Mars and Jupiter, and later moved deeper into the universe.

One night in the late sixties, I was in a neighbor's field putting a battery into my car late at night at about 11:00, when I saw two bright white lights slowly descending from the sky toward me. I noticed there wasn't any sound coming from them, and I started to wonder what they were. They got very low, and passed behind a large hemlock tree, and then were gone. Nothing else was in my line of view for miles; it was all open farmland. My neighbor asked me the next day what I was doing with that old car of mine that would make his TV and radio go on the blink.

I really thought I was cracking up; I'd catch things moving out of the corners of my eyes. I'd see things streaking across the sky very quickly, in broad daylight.

One night at work I saw a man in coveralls, a very tall man with neatly combed hair, walking away from me toward the door. There was a crackling sound coming from him like electricity or static on a radio. He walked through the closed door and disappeared.

I started blanking out on my way to work, not knowing where I was for a few seconds. My body would start to hurt all over. I went to two different doctors for blood work and other tests, which all came out negative. The doctors couldn't find anything wrong with me mentally or physically.

VISITORS IN THE TREES

The event I'm going to relate happened at our house, in a sparsely populated part of the countryside. The house has two stories and is built on a steep hill, so that when you look out the windows on the second floor, you can see into the tops of the trees that grow next to the house.

At about nine PM, what I thought was a large car with bright headlights rolled down our gravel driveway. I was washing the dinner dishes while looking out the kitchen window, which overlooks the circular driveway, but I didn't see any car there, so I thought nothing of it and went on up to bed.

We slept until about seven AM, and when I was back at the kitchen sink after breakfast, I looked out the window to see a woman in a red windbreaker jacket enter the stables. She was wearing white pants, and was holding a long stick in her hand. I told my daughter to go see who it was, but when she returned, she said no one was there. I then saw a man jump off the pump house nearby, which is eight feet tall, and run off toward the trees. He was small, with brown hair, and seemed to bounce in a way that had no relation to gravity.

Next, I left to do some marketing, and when I returned, my husband walked up to me and said, "There are people in the trees! We've been trying to talk to them, but they won't answer." My husband and I walked up to the front porch, where we saw the children calling up at the treetops, "Come down, we won't hurt you." I looked and saw that whoever was up there had constructed some sort of platform.

We went upstairs to look out of the bedroom window at the beings we had been trying to talk to from below. The second-story windows are just at the right height so that we could view them on a level plane. The trees they were in are right outside this window, about sixteen or twenty feet away, so we could see the beings clearly.

I said to my daughter, "Do you see what I see?" She said, "Yes Mom, there's two of them. What's that thing coming out of that one's head?" I noticed that there was some kind of beautiful beaded antenna sticking out of the left side of the head of one of the beings. One of them looked slightly Oriental, and the other seemed more American, but smaller and with a brown mustache. One of them had on a remarkable piece of jewelry—it was a band striped in different metals of all colors: silver, gold, platinum, green, red, purple, and black. We could only see them from the chest up, since the branches and leaves covered the rest of their bodies. I got the feeling they were monitoring everything: our yard, the air, perhaps radio waves.

Since they continued to ignore us, we went downstairs and outside again. I felt brave and wanted to find out more about what was going on, so I went and got our Rottweiler dog. Holding him by the collar, I explored underneath the trees. This caused a commotion, and I saw ten or twelve pairs of legs, all wearing white pants, scamper away from me up the hill. I let go of the dog, and as I approached the house, I saw a woman who was the same type of being you describe in your books. She was dressed in a kelly-green jumpsuit and was too long and thin to be a human being. She was climbing among the branches of one of the trees next to the house. I said, "You have no right to do that without my permission; you should have asked." I instantly realized how silly this accusation sounded, and I quickly opened the front door and went back into the house.

Back upstairs, at the bathroom window this time, I was able to get a closer look at her. She was unlike anything I've ever seen. Her arms were long and unbelievably thin, and she had some sort of faun-colored soft leather flight cap on, of the type that pilots used to wear in the old-fashioned open cockpits of early planes. She was also wearing goggles from the same era, although the lenses were shaped to fit her large, slanted eyes. I can't recall that she had a mouth, or much of a nose. She had soft-looking gloves on, and her jumpsuit was closed down the front with some sort of metal fasteners. She looked like she was engaged in filming, and aimed a black video-type gadget directly at me. I immediately jumped into the shower and jerked the curtain shut as fast as I could, whereupon my husband walked in.

I said, "Did you see her?" and he said no, so determined to learn more, I left the bathroom and went out of the house, across the front yard, and into the rundown paddock area of the property. There I saw the most incredible being that I have ever seen. It was almost indescribable—a silver, crystal, moving mass of energy and light with the exact same striped band of jewelry on it that I described before, perhaps where a neck would be. I stood three feet away from it, awestruck.

The children and I are a little bit vague about what happened during the rest of the day. We remember that at about four PM two of our friends arrived, and I told them to come and see the people in the trees. I insisted that they come upstairs with me and look out the windows, but we saw nothing.

The next morning, as we headed for the beach, we noticed that the birds were back in the trees again. There had been none there the previous day. My daughters and their girlfriend and my husband and I all talked about our adventure while at the beach. Jini had seen the man with the mustache, as well as some of the other beings. My daughters Dana and Ruth verified their height; Jini showed me with her hand how tall they were (about five feet). My husband took the stance of denial, postulating that we'd all somehow had similar hallucinations. Since then, he's acknowledged their presence, but is reluctant to discuss it.

I got very ill after that visit. The following week I lost thirteen pounds. I needed two liters of intravenous fluids the following Friday. On Saturday I was better, after a week of sore throats, fevers, restless nights, and nauseous days. My littlest one has ground her teeth down, and pulls the covers entirely over her head every night, but we're more calm about it now than we used to be.

A LIFETIME OF ENCOUNTERS

The first meeting that you described brought to mind an incident when I was about two years old. Due to my mother's illness, my paternal grandmother took my father, older sister, and me to live in her house. My sister slept in my deceased grandfather's bed. I slept in my sister's bed that had ropes to create a bed rail to confine me. On this day I was standing in bed holding on to the rope when a "playmate" came into the room. The dark figure in your account is a perfect description of the figure that I saw.

During my childhood my family tried to convince me that the beings I saw and played with were figments of my imagination. They teased me about my invisible friends. My sister told her friends who had siblings my age about the imaginary playmates. Since I was the only one to see the creatures, I became the laughingstock of the school. You can gather what the situation did to my self-image and self-esteem of a child growing up in the 1940s when there was no idea of aliens. When I was five and ten years old, I saw child psychiatrists who proclaimed me sane. The beings that I saw looked like the ones in your book. I have felt that one of the visitors was watching over and protecting me since I was five. In my fantasy dreams, I called him my "brother."

My first marriage produced a daughter. I left my first husband when our child was a year old. We moved from St. Louis to Jonesboro, Arkansas, so I could get a divorce. After that, the visits must have increased, as there are more frequent blank spots in my memory. At times, my daughter became difficult for us to wake, and we both had unexplainable problems with our right nostrils. I noticed that the "holes in my memory" seemed to precede periods of unexplainable agitation. The agitation occurred from our move to Memphis to this date. One incident that I recall in Memphis may have some relevance to your work: On a particular morning, my daughter awoke and suddenly announced that she wanted to start a collection of owls. She had previously had no interest in them. At the time I thought it was strange. Your book explained the significance of this statement.

A few months later I became fascinated with the pyramids of Egypt. I looked for a pyramid necklace. A few years later I found a pyramid with a circle inside it, but never did find the one I really wanted. It had the pyramid inside of a square, with the circle inside of it. Whenever I asked for this design I received strange looks. Now I understand the significance of it.

There were several recurrences of a nose problem. Similarly, my husband began to have the same affliction. When his persisted, he went to the doctor. One nostril was an unusual robin's egg blue. The doctor said it was strange and prescribed a medication that worked, much to the doctor's surprise. Did the visitors cause the agitation and the nose problem?

Another incident was when I woke up near dawn dripping wet with perspiration and in an uncontrollable, raw, crude, animal-like terror. In the center of my head I heard a voice say that everything was all right, and to calm down and go to sleep. I took a few deep breaths, laid back down and went to sleep until the alarm woke me. Years later my husband had a similar dream about what he described as a gun pointed just above his eyes on his forehead. He heard a similar voice. On two different occasions I woke in the middle of the night with extreme burning below my ribs and above my diaphragm. Several doses of Pepto-Bismol eased the burning, but not the diarrhea that followed.

Two years ago, the last child married and moved out. That's the next time I had a memory of visitors. My husband had left for work, and I was also preparing to leave. This is the first time I experienced a compelling urge to take a nap. As I was lying in bed, I saw a humanlike figure enter the room. I felt total terror, but I felt as if I knew this entity. The feeling began to calm me down. The being seemed surprised at my reaction. It, or she, asked me why I was afraid, as it had been many years since I'd acted that way. I asked if this had happened to me often, and the answer was "Yes." I was a "chosen" one for them and the Supreme Being. It was then that I made an agreement with the creature. The deal was that it would completely answer all my questions, now and in the future; for this I would cooperate with them totally and completely. With negotiations completed, we proceeded.

The first question I asked was, "How long have you been visiting me?" The reply was both before and after I was in the womb. My sister was born in 1934, and during that pregnancy, my mother suffered with uremic toxemia. The doctor advised that mother have no more children. The entity explained that they had been seeing my mother and monitoring her pregnancy with me. They made tests on me while I was in the womb. The visitor revealed that one of the tests caused my six-week early arrival. It was not her illness, as previously thought. Next I asked the being if they were the things that my aunts told my mother she saw. To explain, mother's sisters decided it was time I knew the truth about my mother. They thought I was old enough to understand, now that I had completed my freshman year at college. They told me about the beings that mother reportedly kept seeing, and the things they allegedly did. Could these events have caused her to lose her sanity and be hospitalized, not brain damage from the toxemia? She kept seeing those things in the hospital until she died, much to the family's dismay. Answering these questions, the being confirmed my suspicions.

One other question I remember asking was why I was told that I'm a "chosen one," and what does that mean? All I remember is something about being trained to do something for the "hierarchy" and the "Supreme One." I will know the rest "when the time comes." Who is the Supreme One? The one that we say is the All-Knowing One God. As far as I know, this was not from my imagination.

When I was reading the section of your book regarding the triad, there was a voice in the center of my head. Quite simply, it said, "The triad is simple. The Saints form one side of the triad; others call them the Ascended Masters, or Spiritual Guides. We, the visitors, form another. When the Masters and the Supreme One came together, we [visitors] were the resistors. When the time comes for all of us to work together, the resistors will be involved in our work together."

What is the goal? The answer continued: "The ultimate aim of the work is to save Earth. This is part of the message that the two of you must work as a team and deliver to the people."

Unlike your experiences, I'm not able to obtain confirmations of mine. Everyone from my childhood is either dead or I've lost contact with them. The exceptions are my sister and two senile aunts. My aunts say that they recall nothing regarding my mother's "hallucinations." My sister says that I am the strangest thing that she recalls from those years. She says she remembers no single incident. None of my children nor my husband recall anything unusual.

A WILD JOY

May 1991: On the night of the full moon at about 3:00 A.M., I went down to the beach, which is opposite where I live. I go there periodically to find some peace. On this occasion there was a fresh wind. This and the effect of the moon were virtually calling us out of the house. I find it very difficult to sleep on such nights.

Spontaneously, I began to run up and down the beach, close to the sea. There was a kind of wild joy in this, as if somehow my running, the crashing waves, and the effect of the wind and moon were all part of one movement. Temporarily I was free from the predominance of rational, discriminative thought.

Then something which I can only describe by a comparison to one of Escher's prints began to happen, a sort of perpetual metamorphosis. I began to see dark, moving shapes, appearing to dance on the water. They seemed to correspond with the dapples on the water itself. At first I doubted what I was beginning to see. A slight shift of focus and these entities would vanish (if that's what they were), and yet with a slight readjustment, a slight relaxation, the forms would be back. The focusing and vividness of these forms, in my perception, corresponded at this point with shifts between doubt and the suspension of my disbelief. This probably took place over few minutes. The doubt or fear of what I was seeing was sufficient to eliminate it, bringing back ordinary "ripples" on an ordinary sea. Yet acceptance brought these dark, dancing forms more into focus. Something else seemed to be involved in this act of perception, a relaxing in my forehead. I allowed whatever was going to happen, happen.

They were like people, but not exactly people. They didn't seem to have a solid shape, but rather seemed to flow from one shape to the next as quickly as swapping partners in a flirtation dance. At one point, they were little black triangles in file, executing impossibly exact figure eights on the sea's surface. Circles and figure eights seemed to be a favorite configuration on this and other occasions. Only a few days later, while observing the sails of the spinnakers, did I realize that the small black triangles were like miniatures by comparison. Then there was something that kept leaping in and out of the water, executing a circle as it did so. It also seemed like they were having a party (!); there was such constant, fast, fluid movement. It also seemed as if this was something I was being allowed to see, like a gift.

As I watched, these moving forms became more and more vivid. They were not very far away from me, but as I was sitting down they were somewhat higher. I began to get a bit nervous, and eventually stood up and left.

For over what was about a month, I'd occasionally go down to the beach at different times of the day and night, in the hope/fear of seeing "them," the "elementals," as I named them, thinking that was probably what they were, if they existed. Sometimes during the day I'd spot them without really meaning to, even when children and grandparents were playing in the water. They were never in sharp focus, and like the waves themselves, they were always in a flurry. I could never be sure that what I was seeing, or not seeing, was real.

One morning shortly after dawn, I walked down to the beach having had no sleep (though not because of this subject). I was feeling as one sometimes does in such circumstances, pleasantly drained, not only of energy but of excess mental activity. Sure enough, at a glance, there they were on the left side of the mole. They seemed to be schooling around the tip to the other side, though anyone else would have seen, I'm sure, just some fairly normal current activity. These beings were never quite as in focus or as frightening as on that first night, when I expected nothing. I walked onto the mole and up for a closer look. They seemed to be heading toward the next mole, near the yacht club. As I left the mole and walked hurriedly along the beach, it occurred to me that I was acting like a foolish child, "chasing faeries." It also occurred to me that maybe I was being fooled by some trick of the light at that hour. Sure enough, the shoal seemed to recede from the shoreline as I reached the spot where I would have been able to get a clearer view.

I walked back along the beach a little frustrated, caught

between curiosity and a sense of my own foolishness. Stopping to look out to sea for any change on the conditions there, I saw what appeared to be a tribe of little blue men. They were the color of the sea and were standing on the water (hence, barely visible), several yards out and facing me in a slightly concave formation. They were holding spears, or possibly long staffs. I observed them briefly, then walked on.

I had grown weary and I resolved, after that episode, not to go on seeking out these phenomenon. The act of doing so was taking me away from my center, drawing me off balance. I did not feel threatened by these blue people, but retrospectively I think that their stance was a "warding off," telling me to mind my own business and to leave them to get on with theirs.

"LET US BEGIN"

I have just recently, in fact, about a week ago, spoken my first words to another human being regarding these fascinating memories and experiences.

What could possibly explain what has been happening over the years—physical contact with thin, gray, short beings, sometimes in my room, sometimes in dark, drab rooms foreign in appearance to anything I have seen elsewhere, sometimes in open, stone-hewn buildings where I am sent with others to learn, but almost always utterly terrifying.

I have had memories for years of several different sightings of UFOs, which I have remembered ever since they occurred. My first memory of sighting was when I was only about three years old, in the summer of 1964. I remember looking at "the funny moon" with one older sister and two older brothers, gazing out of the window in my brothers' room. I remember it dancing in the distance, a silver-gray, iridescent disk, over the woods that used to stand, at that time, behind my house. We were frightened and mesmerized at the same time. I recall seeing something that looked like a rooftop-type TV antenna sticking out of the bottom of it. We giggled as we watched it move up and down, then shoot toward the city, but then we all screamed as it darted close to the house, and then the memory ends.

My next sighting was much later, when I was thirteen, in 1974. I was going to bed one night and my mother and sister were downstairs. I saw, about a mile or so off, a silver-gray, iridescent disk, which seemed to be wobbling on its vertical axis as it stood otherwise motionless in the air. I was filled with terror and plastered myself against the bedroom wall beside the window so "they" couldn't see me. I wasn't consciously aware of just who "they" might be at the time, but I felt certain that they knew I had seen them. I finally worked up the courage to peel myself off the wall, get off the bed, and run out of my room and down the steps to my sister and mother to tell them that we needed to run and hide. I went in, and my mother sent me to bed. I lay awake for some time, praying.

The next sighting was not too long afterward. My mother and sister and I were driving home in the evening in a deserted, forested part of the state. I decided to lay down on the backseat for a while, and look up at the nighttime sky out of the rear window. I noticed a star that seemed to be pulsating. It started to move, slowly at first, then faster. It shot across the freeway behind us some distance away, then doubled back, and came much, much closer. It was cigar-shaped, and seemed almost split down the middle, color-wise. One half was a bright, intense kelly green, and the other, scarlet red. The next thing I remember, it was a little after one o'clock in the morning, and we were sitting at home at our dining room table. We still had our jackets on, sitting around the table, saying nothing. My mother asked me to check if she put the car in the garage or not. It was there. Nothing else was said between us. I was extremely tired, and went straight to bed and fell immediately asleep. We never spoke of the incident.

I was in my bunk bed one night thinking. I thought I was awake. I saw a small ball of light traveling across the floor, and thought that it must have been a small glow-in-the-dark Superball. The light was whitish yellow. It went under the bed, and I went under the bed to retrieve it. In my dream, I could even hear my sister breathing as she slept in the bunk over me. As I crawled under the bed, I was no longer under there, but instead in a dank, gray-brown cavern. Many people were being led to their fate by skeletons, with large, slanted, glowing eyes. The ball of light I had followed then drifted past me, and entered the eye of one of the skeletons. It turned to face me, hissed loudly, and blew a kind of smoke at me.

Our "friend in the basement" used to make things fly off the back of the kitchen table when we would walk past, make the bread loaf fly off the top of the oven at us, terrify pets, and unnerve us kids. My mother insisted that it was nothing to be afraid of, that it was completely harmless. Activity regarding "the friend" settled down over the years, as my sibs moved out, and our daily schedules picked up. But during this time of meditation and development, I changed. I started having nightmares of the thing, and could only see large, black slanted eyes in the basement of my dreams. I find this interesting since at this point in time, there were no pictures of visitors as we see them today. One time in particular was exceptionally frightening. I was lying down in my bed, writing in my journal. I felt the presence of the thing in the room with me, and I was terrified. I tried to run, to get to the garage to my bike. Upon reaching my bike, the thing came up on my back, and I took off into flight—riding my bike through the air, struggling in vain to escape, but instead being overtaken. The last thing I remember is its voice coming over my vocal cords, from within me, and I was totally not in control. It said, "Let us begin."

Something extraordinary happened to my son Kevin (not his real name) a few months ago. The boys had said they'd seen the little people in their room. One morning, after one such night, Kevin woke up with a triangular-shaped mark on his left temple, similar to a scrape or brush burn mark, except that it was a perfect little isosceles triangle. When I asked him how he had gotten the mark, he said he had no idea, and didn't know it was there.

My memories of actual events upon visitation are very fragmented. I have glimpses of a dark room, with a very strange smell, almost like moist dirt. The lights are very dim. I remember lying on a metal table, but it was warm, instead of cold. I remember the visitors, I remember me screaming. I remember one being, in particular, whom I perceived as a male, who was always there. I knew him, and he was the one who could always calm me. I remember being reassured by his presence. The eyes were crazy—almost black, shiny, but they didn't seem moist to me. They had almost a fine texture to their surface, like the surface of a fly's eyes might appear under a microscope. The face reminded me of a praying mantis. Their bodies reminded me of a child's but thinner, around four to five feet in height. I don't remember what the clothes were like. The touch of their skin reminded me of soft glove leather, but again, it didn't seem as moist as ours. Some had little pot bellies, with real thin arms and legs.

Every time this happens, I wake up with a migraine and a nasty allergy attack that lasts for several days. I have one clear memory of lying back on a table, naked, but warm. There were three grays around me, but I had the impression there were many more in the background. There was one of them, very tall and mean looking, and I thought at the time he was the leader, but he impressed me as being almost mechanical. They had two rods shaped like long drumsticks, which appeared to be made of some copper alloy, which they held in their hands. They were pressing the ends of the rods at various places on my body, and sound was emitted of varying pitch, like a tone, when they pressed them. The tone changed as they moved them to different points on my body. Each time the sound was emitted they looked over at what I perceived to be a screen of some type. I remember a metal object shaped roughly like a microphone being inserted up my rectum. I remember the male telling me that it would be okay, that they wouldn't hurt me. But they did. I feel at times they were surprised by the presence of pain. I told them I wanted to go home, and asked why were they doing this to me? They told me I was one of the chosen ones. At the time, it made sense to me.

ACROSS THE GENERATIONS

Let me give you a brief history of myself and my family. We suspect my grandfather on my mom's side of the family had encounters. He loved the sky and the stars and would always take his family out on summer nights and teach them the constellations. He passed on in 1976, so we are unable to ask him about anything. My uncle, though, found some old papers that he had written when he was in elementary school. They were about him and his dog and how they went out in the woods behind his house and a UFO would pick him up and he would go traveling.

My mom remembers seeing an orange light coming toward her through the living-room window above the trees when she was five, which was in 1945. There was nothing at that time to in any way give her ideas about UFOs and it makes me mad how the debunkers say that all of this is just media oriented. She also has memories of many encounters that she did not need hypnosis to bring up.

She never told anyone, though, until she was forty-eight years old. That was when I brought home a magazine article that I had read in a doctor's office. It was about the abduction phenomena. I had never heard of it, or if I had, never paid any attention to it. After reading this article I felt I was involved. Mom at that time opened up after forty-three years (after her first experience) and told me about her experiences. I still stuffed everything into the mind closet though until I had a near death experience in 1988. After that the flood gates opened. I began to have anxiety attacks like you would not believe. I was so afraid to go to sleep at night. I slept with the lights on and the shotgun loaded. I am a registered nurse and was a Baptist and this was so unlike me. After the near death experience I came back believing in reincarnation. That does not float well with Baptists, so I left the church. I feel that I am more spiritual, though, today than I ever was while in the church.

My son and I both began having nosebleeds. Often, his battery operated toys would come on in the middle of the night and I would hear footsteps in the hall. I became a basket case. I became interested in Native American healing and started wearing a medicine bag around my neck at night, which I thought would ward off the attacks but to no avail. I woke up one morning with my medicine bag stuffed in my underwear. Another morning I woke up and must have had blood drawn out of my jugular vein, which left a needle mark and a bruise for a week. That made me angry; being a nurse I know they could have at least held pressure long enough so I would not bruise. When I remarried, they started in on my husband. He actually saw a small gray walking down the hall one night leaving my son's room. It turned and looked at him, and as soon as he saw my husband looking at him, my husband said it turned into a red tracer and was gone. My husband was up with the gun and he searched the house, but there was nothing there, of course. He shook for hours though.

Earlier that year, in the winter, I was dreaming I was with these Indian women and all of a sudden this beautiful Indian woman appeared and told me I had to pray and she led me in the most beautiful prayer I have ever heard and I woke up crying. At the same time I woke up, my son was knocking on the door asking if he could come in. I let him crawl in bed with me and he was ice cold. I turned the lights on and his lips were blue. He was shaking and then started vomiting. I know he had been outside, but he didn't leave or come in through a door. He recuperated, and I barely did. I don't know who that woman was, maybe my spirit guide, but I feel she saved my son's life. Later on that summer I had a nervous breakdown. My panic attacks had become so severe that I was afraid to leave the house and afraid to stay at home alone. I checked myself into a local mental health center. I was also depressed.

After five days there I mentioned to my psychiatrist about my encounters. I was so afraid that she would throw away the key and I would be stuck there from then on. But she said she believed that there was a lot more to life than just what we could see or feel in our reality. So she hypnotized me. I had to know if they were real or if I was really just crazy. The first time she just used a light hypnosis to get me used to it and the next time put me under deeper. Lo and behold, there they were. She said nothing leading to me, but I was back in my bed as a little girl, terrified. And then he came. He took me out my window and toward a gray object in the sky. Everything else was blocked until I could remember being back in my bed and this little whitish gray creature with huge black eyes, that looked like oil spills that were so deep that they went on forever, standing there. But what was so amazing about it was that this little person tucked me into bed. Pulled the covers up over me and patted them and me and stayed. It was almost like he smiled at me. Then he was gone. But I woke up from hypnosis elated. For one, I wasn't crazy, and for the other, this being seemed to actually care for me. It changed my whole perspective. I have done some other hypnosis but a lot of things are remembered now without it. When you mentioned cinnamon smell I almost flipped out. I have never read that in any book, but my son on many occasions has mentioned cinnamon. He is eleven now. One time the beings came and sprinkled cinnamon on him so that he could fly and he says he did.

Another time he was with other kids at these waterfalls and the one he slid down was a cinnamon one, and then at the bottom he was taken into what he thought was a cave, and one of the small gray beings who was sitting at this huge computer console motioned for him to come over. Tim (not his real name) asked him why they were coming and bothering us, and he said that the little being told him why but when he woke up he could not remember. I would love to know what the answer he gave to him was, but I am not going to have him hypnotized. He still does not like to sleep alone at night, but as I have accepted everything better, so has he. He knows he can come to me and talk about his experiences, which have been many, including a black helicopter chasing him when he was in a craft. But we don't talk about UFO stuff unless he brings it up or if we see something on TV. His fear is lessening. All I can do is assure him that I, and his grandmother, have always come back and nothing bad has ever happened to us.

They have shown me portions of the future and I think

they took me to a civilized planet one time. It could have just been a future earth, but it was wonderful. When I woke up, or came back, I hand wrote ten pages with pictures of the place I had been. All of the houses were round, and there were no policemen. There was no money; their planet was totally clean, with nourishing the soul and the earth as top priority. Even if it was just a dream, it was a good one. When I was in the craft there was no sensation of movement, as you also described.

Well, we know our experiences are real and everyone else will too, someday. Then everyone is going to be beating our doors down, wanting answers, when now all they can do is laugh. Maybe it is true, he who laughs last, laughs best. I could go on and on about my family's experiences.

ENCOUNTERS WITH A MARTIAN (?)

Early in 1987, I was living in New England with my husband and infant son. One night in late January after a very heavy snowstorm, I had an experience perhaps familiar to you. In the middle of the night, I had the sensation of being wide awake but my body was relaxed, or under someone else's control. A voice or some words came into my head, like a ticker-tape machine or an electronic billboard in Times Square. It said to me, "Be calm. Relax, now. There's nothing to fear; everything is under control." Of course I panicked, and could feel my muscles constrict with tension. This action lessened the outside control over my body, and I began to turn my head to the left. The image/voice told me not to do so, nearly implored me not to do this. I had to see this being.

I was startled to see a very slim creature about four and a half feet tall. On its head it wore a WWI-style pie plate helmet, along with a large trench coat or lab coat with padded shoulders, a thin and tight bodysuit or jumpsuit, a belt with odd-looking tools on it, and an X-shaped escutcheon on its chest. It had a very large head, no ears, a tiny mouth and enormous eyes. There was a power emanating from those eyes that seemed to fetter me into myself. The eyes also seemed to glow.

I turned my head to the right because I feared that something was being done to my husband. He lay in such a deep sleep that he seemed dead. Then I saw his chest move and noticed a cool glow surrounding him. Above him floated a luminous blue spheroid, inside of which floated a spiral-shaped thing, like an embryo. This spheroid was about five inches in diameter, connected to him by a pellucid silvery cord, and was suspended in midair about six inches above his groin. There were tiny, sparkling bits of colored light in a sort of liquid, moving about within the spheroid. I asked what was happening to my son Malcolm, and the voice answered that he was fine, and untouched. While all of this was taking place, I had the sensation of swimming or lying upon a water bed, of floating. After asking about my son I said, "Well, I've seen enough," and fell into a deep sleep. Although there was yet the feeling of buoyancy, another part of me was drawn exponentially downward, into an unfathomable morass of quiet and calm.

When next I woke, the brilliant cold light of a winter morning spilled across the floor in jagged icicles. "How black it was last night," I thought, and suddenly the events of the previous night streamed before me. Fragments of past memories taunted me, reminding me that this had happened before. Not wholly without rancor, I realized that this recent visit was but one of many. There was a faint trace of something bitter on my tongue. I left the room, barefoot across the frigid floor, and went into my son's room. He was then eighteen months old, beefy and happy, and my joy. Carefully lifting him out of his crib I turned him over and over again. I felt like some worried dog, sniffing out a patch of bare earth for a lost bone. Holding my son before me, I looked up his nose and in his ears, although I didn't know why. Afterwards I hugged him close, breathing in his pungent warmth. My husband called up a morning greeting from downstairs. Taking my son with me, I walked down to our kitchen to cook breakfast. Over our meal my husband made a comment about what a heavy but tiring sleep he'd had that night before. I decided to broach the subject of our "visit." I prefaced this by saying, "You may think I'm crazy, but—" He listened patiently, without surprise or disbelief. Afterward he said that what had happened was probably not a nightmare, and that my fear was completely natural. He added that I shouldn't allow my anxiety to swallow me up.

The next couple of days were a trial of my patience and tenacity. I half expected to walk through walls one minute, and blind myself to insanity the next. Roller coasters have never been my favored mode of transportation. About four or five days after the "'visit" you were on TV, talking about your book Communion. I realized, upon watching your interview, that I wasn't alone in all of this. Although my fears lessened somewhat, a strong cord of compassion grew within me outward to you. On the next day came our weekly appointment with our therapist, who seemed not at all startled or affronted at my news. In fact, she was so blasé and cognizant of visitor contact that I've wondered if she wasn't one of them. She was of great help to me.

Since then, we've moved back to California, my home state. The visitor contact has continued, and I have read your two books on the subject. Like yourself, I have strong and vivid memories of infancy and of early childhood contact with the visitors. Other members of my family were infrequently involved, although it appears that I was the focus of contact. I've learned to meditate, using a method taught by the Cherokee Indians. Meditation has helped me to cope as well as to maintain a less traumatic form of communication with the visitors.

I hope you feel the same sense of discharge and awe that I feel, and that there's a thread of common experience connecting us all.

VISITOR IN A BLACK SEDAN

In October of 1979, I picked up my girlfriend (who is now my wife) around midnight. We were going parking, like young people do. We lived in a rural farm community and drove out to the deserted part of a dead-end road. It was a gravel road with only two houses on it, and the closest house was approximately two hundred feet away. There was some heat lightning going on nearby.

We were parked for a little over an hour, when I happened to look out the window on the driver's side and noticed that a car was parked facing in our direction. It appeared to be an old 1930s model automobile, either black or gray in color. This was quite spooky, because it had to have arrived in total silence, since we didn't hear a sound. Then we noticed a small child, no more than four feet tall, standing beside the car. We were terrified, because we didn't believe that a child would be out alone at that time of the morning.

It just stood there staring at us, holding a metallic object like a pail. It was wearing what appeared to be a full head-to-toe child's pajama suit with footies attached. Let me tell you, it had an unearthly look about it.

Our first impression of this creepy little being was that we must have been witnessing the ghost of a child that was killed in an auto accident. There was no moon out that night, but we had the impression of a light up above us. The being walked across in front of the strange car, causing the pail to swing back and forth with a pendulum-like motion. He went along to the passenger side and just jumped in. There was no door that swung open, as there would have been on an ordinary car.

Shortly after he went into the car, he reappeared standing on the driver's side again, where we had originally seen him. During this time, we kept commenting to each other about how could this possibly be happening. At this point, we were not only extremely frightened, we were in a state of shock. The next thing we knew, the car was gone, alien and all. I don't recall how long the overhead light source stayed after the car left, but we both feel it remained for a short time after it was gone. We wondered: could the visitors be using dummy cars in order to get closer to people?

High Pressure

I wish I knew how to start this letter. I feel like a complete idiot. It's not as if I really think what happened to me has anything to do with "visitors." I didn't see any little people or any bright lights in the sky. What happened to me, though I can't explain it, would not have moved me to write to you if it hadn't been for one particular aspect of the experience. I've recently been reading your new book, *Breakthrough*, and you again discuss the episode you experienced when you heard knocking in three sets of three. After my experience, which took place three years ago, I thought of a book I had read a couple of years previously (I think it was *Transformation*) and how in it you had related the knocking incident. I wondered if there was any connection; I even considered writing to you then, but I talked myself out of it. After all, I didn't hear any knocking, and the sequence of three times three that I experienced was probably just a coincidence. But, on the other hand, I'm mistrustful of coincidence, so here's what happened.

At that time, May 1992, I lived only a few minutes from work and it was my habit to go home for lunch. I don't remember anything at all about that lunch hour except that, shortly before it was time to go back to work, I laid down on my bed. After only a moment or so I suddenly (and without even realizing what I was about to do so) reached over to my bedside table and picked up the receiver of the phone. I very calmly dialed my work number and spoke to a coworker, telling her I had a terrible headache and would not be coming back that day. Then I hung up. The calmness with which I had performed this act broke instantly. I was appalled at what I had done. I felt fine! This was completely out of character for me and I had no idea why I had done it. I began to berate myself. I can remember exactly what I said to myself. I was very much alert, very much awake. Then suddenly I fell asleep. I went from wide awake to sound asleep in a few seconds.

I don't know how long I was asleep. I only know that when

the experience I'm about to relate began, I was no longer asleep. It seemed completely real and, as far as I am able to judge, it happened while I was wide awake.

As I lay on the bed sleeping, I was suddenly awakened by some kind of "force" which "hit" me right in the center of my chest. I could see nothing, but the quality of the light in the room seemed odd, as if I were looking into something opaque like a heavy mist or fog. The pressure was incredible! It pushed me back into the mattress with such force that I thought the bed would break. I was panic-stricken. I felt I couldn't breathe, that this pressure had forced all the air out of my lungs.

But just as I thought I might die, it suddenly became very clear to me that whatever was doing this was not trying to hurt me. My first instinct had been that I was being attacked. Now I realized that this was not the case. I told myself that if I would just try to calm down I would be able to breathe. I told myself it would be all right. (I never heard any voice speaking to me except my own.) Soon I found, although the pressure had not decreased, I could breathe. I was still frightened, but I was calmer. There was no pain, just incredible pressure. As the pressure continued, I realized that my feet were slowly floating (or being drawn) up toward the ceiling. This continued until only my head and the tops of my shoulders were touching the mattress. At that point, my feet came slowly back down until I was once again lying on the mattress. The pressure stopped, but began again immediately and the whole process was repeated—pressure pushing me down into the mattress, then feet floating up toward the ceiling until I was almost standing on my head. This happened *three times* and then stopped. It stopped very suddenly and I lay there for a few seconds with my mind racing. I remember thinking, "I've got to do something! I must be having a seizure or something! I've got to call for help!" As that last thought occurred to me, I reached desperately for the phone by my bed, but in the same instant, the force hit my chest again and the whole series of three started again. And then again. Three times three.

After the third series of three, I drifted back to sleep as if nothing had happened and began dreaming. I dreamed that I wanted to call for help and that this time I was able to reach the phone. I called my parents' house and my father answered. I said, "Dad, it's Marianne." He said "Who?" I said, "Your daughter, Marianne." There was a long pause. Finally, I said, "Listen, Dad, is Mom there? This is important." I listened as he spoke to my mother, who must have been in another room. He said, "There's someone on the phone who claims to be Marianne. Do you want to speak to her?"

At that point I woke up abruptly. I leapt up from the bed and immediately turned around to look down at the spot where I had been lying. I thought, "I'll never be able to lie down on that bed again!" I paced around the bedroom for awhile trying to decide what to do. I was no longer calm. I was close to hysteria. I remember sitting on the stairs that led up to my loft bedroom, hugging my knees and trying not to start screaming. Finally, I went downstairs and called a good friend at work. I told her not to ask me any questions, but to just talk to me about normal things, anything, just so it was normal, everyday stuff. Thank God for Susan. She tried to do as I asked, and every so often she'd say, "Are you okay now? Can you tell me what happened now?" Finally, I was able to relate to her what had happened. Obviously, neither of us knew what to make of it. I also told her about the dream and she asked me what I thought it meant. Without thinking, I said, "I believe it means that something has happened to me—something that changed me so profoundly that even my own parents didn't recognize me."

Looking back on this experience now, I have tried to remember exactly how the room seemed as it was occurring. As I said before, I had the impression that there was an oddness to the light in the room. I couldn't really see anything at all, but I know it wasn't dark. Also, I seem to remember a loud, roaring sound as the pressure was being exerted on my chest, but frankly, I suspect both of these effects may have been due to my state of panic. Aside from that, I don't think there is anything more I can add concerning the experience.

This might make more sense (at least to you) if I had a history of seeing fireballs or owls that turn into little beings, but I don't. The only experience I've ever had that had anything to do with the "visitors" happened in an optometrist's office. No, I'm not kidding, I was having a routine eye exam. Little gray beings with big black eyes were the farthest thing from my mind (whether or not I'd need new glasses was the closest). Then the doctor shone a bright light in my right eye and as he did so I saw a picture, crystal clear, directly in front of me. It was the head and shoulders of a creature such as is depicted on the cover of *Communion*. It was not exactly the same—for one thing it was much more real—not a drawing, but more like a photograph. Also, the face was not so long and narrow and the skin was a lighter color. I was so startled that at first I didn't hear the doctor's instructions as he prepared to go on with his exam and he had to repeat them.

Okay, as I was writing the previous paragraph I remembered something else that probably means nothing at all. This must have been shortly after *Communion* was published. I was standing at the checkout of my favorite Walden Books. The cashier was already ringing up my purchases when I turned to look to my right and caught sight of the drawing on the cover of *Communion*. With the cashier telling me how much I owed, and with people waiting in line, I simply walked away toward the book, picked it up, brought it back to the cashier and said, "I think I have to have this." I had never heard of you, never heard of alien abductions and, under the circumstances, I have always felt that my behavior that day was more than a little peculiar.

One more thing, and, again, this probably means nothing. Something you mentioned in *Transformation* gave me quite a turn when I read it. It was something about fireballs and black sedans being common memories of people who have had the abduction experience. It was the part about the black sedans. When I was five years old, I was walking on a street near my home. In those days (thirty-seven years ago) this was a very quiet area, very little traffic. With me was a four-year-old playmate named Sally. We were nearing the corner when a big black sedan came swooping around the corner and right up on the sidewalk heading straight for us. I grabbed Sally and pulled her with me into some bushes at the side of the road. Later I remember telling my mother that a big, black car tried to run over us. Now I would say that this is not a big deal and was probably exactly what it appeared to be except for a couple of things. First, for me, this is the memory that wouldn't die. I've questioned my mother any number of times about what she remembers about it, whether she believed me at the time, what state my clothes were in when I came to tell her, if I looked like I'd been jumping into bushes, if it was likely that a thing like that would have happened in such a quiet area, and on and on. It has bordered on obsession over the years. Not long ago, while going through some boxes in the attic of my parents' house, I came across a short story I had apparently started and then given up on while I was in junior high. The name of the story was "The Black Sedan." So, after reading *Transformation*, I wondered. Now, Sally (who is still a close friend) doesn't remember the incident at all.

But, Jane is the fireball queen! They come careening through her bedroom at night. There was one instance of her seeing a fireball in her bedroom that also coincided with her seeing her brother, who was killed in Vietnam. During that same episode she remembers an out-of-body experience and later running through the house convinced that someone was trying to take away her children, who were small at the time. You'll just have to take my word for it—Jane is not a loon. She has also related to me a vivid dream she had when she was a teenager. She had felt compelled to lie down in the middle of the day and had immediately fallen asleep. She then saw a hand holding some sort of wand which proceeded to point to a map of stars and planets. This dream made a tremendous impression on her, and at the time she told me about it, she had not read any of your books.

ARE WE READY YET?

In December of 1989, I was on a business trip to Florida, at Disney World to be exact, staying at the Contemporary Hotel. I'd landed that morning and had worked for about seventeen straight hours. I arrived back in my room at approximately 12:00 midnight, exhausted, and my back proceeded to go into a spasm like I'd never experienced before.

In an effort to get some relief, I took all the sheets/blankets onto the floor at the foot of the bed. I started doing some deep-breathing exercises lying face up, with my feet facing the sliding glass doors and my head toward the entrance to the room. I have no recollection of falling asleep; to my mind I was still awake and in pain. Something caused me to raise up on my elbow and turn back toward the door of the room.

To my utter amazement, there stood a form about three to four feet high in the doorway of my room. My best description is that it was all energy and light! It was something like the shape of an elongated light bulb. I don't recall anything like arms or legs but it did have a face, although as in one description in your book, it only had the hint of eyes, nose, and a mouth. It more closely resembled an embryo than a fully formed being.

Fear immediately took over. I felt its all-compelling power from the top of my head to the tips of my toes! It was the kind of fear you so perfectly described in your books. I've a sense that I was standing at this point, but I don't recall standing up. At the same time, I think I was still lying down. What I do know for sure is that I was paralyzed; I couldn't move or make a sound.

Within seconds I was told telepathically, "I am not here to hurt you. I am here to tell you something." Immediately, the fear subsided and I answered telepathically, "I am not ready to hear it." At this point I did something that I've regarded as very strange until I read your book; I rolled over and immediately went back to sleep!

When I awoke the next morning, I remembered the incident

very clearly. Of course I tried very hard to tell myself it was just a very vivid nightmare. However, my problem was that when I have a nightmare I never roll over and go back to sleep. I usually lie awake for hours with my heart beating wildly. Also, I've never had a nightmare that took place in the surroundings that I was in, complete with the position I was in on the floor and the colors, etc. that were in that room.

On returning home to Connecticut the next day, I told my husband and a few close friends about it. They had some good laughs about "Dopey" from *Snow White* coming to visit me in Disney World. Then one friend who's into metaphysics told me that I must read *Communion*. It took me over a year to get the courage to pick it up. Interestingly, I'd vaguely heard of the book, but had never seen it in a store, and had never seen the cover. A few days after being told to read it, I was in a used bookstore with a friend. We wandered off in different directions until she heard me let out a small scream. Your book *Communion* had fallen off the shelf at my feet. Since then, every time I went into a bookstore that was the first thing I'd see. Yet it took me, as I said, over a year to actually buy and read it. Of course, my first reaction after reading it was to wonder if, in fact, what I recalled was all that had taken place the night of my experience.

The only other experience I've had was during last spring. A few of my friends were picking me up for a meeting. When they came into the house they jokingly said, "There's a UFO over your house." We left and proceeded up Route 7 toward Danbury. In a very short time we saw what they said they'd seen over my house. It appeared to be hanging just over the treetops. It had kind of an oblong shape, like a bus, and very brilliant white lights. It moved very slowly, and we followed it up Route 7, stopping when it stopped and then following it again. When we got as far as the Danbury Airport, we lost sight of it.

The next day there were reports of a UFO sighting on Route 84, which is just above Danbury Airport. Of course, it was followed by a report that it was something being tested at the airport. I also know of someone who had said that one came down very low in their backyard nearby.

A Lesson from the Dark

My younger brother owns a recording studio, and as I am an amateur pianist, arranger, and composer, I frequently use the studio to make demonstration tapes. Even a three-minute recording can take several days to put together, and on this occasion I stayed overnight the first night, finished recording about midnight, then headed home on the sixty-mile drive back to my place.

When I got home, it was about 1:45 A.M., and I undressed and slipped into bed without waking my wife. I rested my head on the pillow, feeling very happy about the results of the recording session and looking forward to playing the results for my family on the stereo the following morning.

I reckon my head had been on the pillow for less than thirty seconds when, for want of a better word, it exploded—the only way I can describe that shocking sensation is that I thought a bomb had blown me to pieces and that "I" was nowhere and had ceased to be. After a few seconds, the vacuum of what used to be me was filled by an entity of total evil. This evil thing so terrified me that I wanted to start fighting, until I became conscious that I was unable to move my limbs. Although I was screaming to my wife to wake up and help me, my lips barely moved and the screams were whispers. Eventually, maybe a minute or two later, my bodily control returned and with it real screams. Sally awoke immediately and did her best to comfort me. I realized that my pillow was wet with tears.

After I gained control, I explained to Sally what had happened. She switched on the bedroom light and made us some coffee, and we sat up in bed talking quietly about what might have occurred. I became calmer. I knew the whole thing had not been a dream, because I'd never fallen asleep. We both put it down to a rather violent out-of-body experience, perhaps even an attempted "possession" by some entity. We were alarmed, but everything was okay, and since Sally had to go to work in the morning and I had the housework and paintings to paint, we had decided to turn off the light and get some sleep, when we both said, "What's that noise?"

We heard a low humming sound. A quick glance at the clock told us it was 2:30 A.M. The humming soon changed to a deep, fast throbbing. It didn't sound like a plane, or a truck, or a car. It got louder and stopped right over the roof of the house, directly above our bedroom. There were no flashing or glowing lights, just a very loud thumping sound right over our double bed. We froze; what on earth could be on the roof?

Before I could volunteer to go outside and see what was going on, all our nice little stories about OBEs and possession were shattered, as something invisible grabbed me by the chest and started pulling with amazing force. I felt like my soul, not my body, was being pulled up vertically toward the loud throbbing noise, and although I thought it would be futile, I screamed for Sally to lie on top of me. When she did this, the sensation of pulling eased a little.

I was screaming and struggling against an invisible "beam," with my wife lying on top of me in my bed at 2:30 A.M. What a sight we would have presented if someone had walked in! It might have seemed funny later, if it hadn't gone on for another two hours. Throughout the night the children, who slept directly across the hall from us, never woke up. "They" pulled, I resisted, Sally hung on and the engine throbbed, until finally it went away. Exhausted and badly shaken, we fell asleep.

A POLICEMAN'S STRANGE JOURNEY

I have debated with myself for a week over whether or not I should write to you. I am doing so out of frustration, and I do so with great reservation. I considered that you might throw this correspondence away, thinking that I was some kind of quack.

I have never spoken with anyone about certain things in my life. What brought me to you was your books. I have read them all, including *Breakthrough*, which I just now finished. Frankly, the books scare the hell out of me. I did not sleep well for weeks following *Communion*. I again feel very restless after reading *Breakthrough*. I cannot explain this. Tell me I am imagining things.

First of all, please allow me to provide you with some personal background information. I am a forty-eight-year-old police detective sergeant, with twenty-two years of service. I was a SWAT Team Commander for nine of those twenty-two years, and I can assure you that I am not prone to hysteria in a crisis. Prior to my police service, I served two years active duty with the United States Army and four years reserve duty. I am divorced and have two grown children, a daughter, twenty-three, and a son who is twenty-two and married. I have a girlfriend in her forties who is a very steady person and works as a paralegal. She is the most honest and trustworthy person I know, and I love her dearly.

I decided to finally talk to someone, after an incident occurred on May 18, 1996, at 12:35 P.M. I live near a farm market and shop there often. The trip to the market usually takes about twenty minutes. I was out this day, running errands and so forth, with Michele. We decided to go to the market, as we usually did, on weekends. I was traveling westbound on Route 40 toward the Delaware Memorial Bridge. About two miles before the bridge, my girlfriend looked up through the windshield, to my left, with a puzzled look on her face. After about five seconds, she said to me, "What in the hell is that!" I could not see anything from my view angle, and ducked my head to see upward between the steering wheel and the top of my windshield. What I then saw was what appeared to be the trailing edge of a structure similar to a fuselage. My view of the object lasted no longer than one second, maybe two, because the object had disappeared over my vehicle, out of my viewing area.

My girlfriend scrambled around in the seat in an attempt to look out of the rear window of the car, and I hurriedly pulled to the shoulder of the road, to get a better look. It may have taken me five seconds to stop. We both got out of the truck and looked in the direction the object was traveling, but saw nothing. We both stood there and looked at each other like we were nuts. I then handed her a piece of paper and told her to mark the time on it. I then asked her to draw the object that she saw, as she got a full look at it and I did not. She took the paper and my pen and drew the object in about twenty seconds, without hesitation. This rather surprised me because I watched her as she drew it, and she never even stopped to think about what she was going to draw. She just did it. When she handed it to me, it felt like every hair on my body stood up. I then told her to mark the time of the drawing. My watch said 12:38 P.M., three minutes after the sighting.

We both drove away feeling very "displaced," as you do when you first wake up. My stomach was now cramped and I felt like I had to use the bathroom immediately. I suggested that being we were so close to home, maybe I should go there and relieve my sudden stomach problem. My return home was accomplished in about fifteen minutes I estimate. En route to my home, I contacted my police department by mobile telephone in an attempt to ascertain whether or not anyone else had seen what we saw and reported it. I did not tell the dispatcher what we saw. I merely asked him if anyone had called in complaining about low-flying aircraft. It's a rural area and crop dusters are not unusual. The dispatcher said that he had received no calls of that type. We arrived home, where we spent about ten minutes and left again.

En route to our destination, I asked my girlfriend to describe what she saw, because I did not want to tell her my observations and risk influencing her version of what happened. She told me that as we were driving along, a black object appeared in the sky, just above and left forward of my vehicle. She told me that it sort of blossomed into view from out of nowhere. "Not real fast and not real slow," she said.

The object was moving from southwest to northeast, at what she described as "moderate" speed. Not as slow as a jet liner in the sky, but not as fast as a fighter, either. She described the object as elongated and black with a little brownish color on the trailing edges. The object was not shiny, but rather dusty looking, in that no sunlight reflected off it. It appeared to be at about 250 feet altitude. Michele described its size as about the same or larger than a car. The description concurred with my brief observation of the object. I did not see it in its entirety, but I did see part of the left rear of the object and saw the same colors as Michele.

Michele pulled out the drawing she had completed and looked at it questioningly. I asked her what she was thinking about. She pointed to a tail structure she had drawn and asked me what it was for. By her description of the structure, I could not explain it. In the past, I accumulated about eighty hours on a student pilot license and I consider myself adequately schooled in weather and related subjects. This was no weather phenomenon. We terminated our conversation on the subject, for the time being, as we arrived at our destination.

The parking lot to the market was full so I asked my girlfriend to drive around the lot while I went inside to get a pack of small cigars. She agreed to do so and my trip inside the market took less than ten minutes. We then started for home. As we drove over the bridge, I placed a call to her daughter, who was still at my house, and asked her to turn the television on and get the weather data from our local weather station. I recall the time now being around 2:38 P.M. When I commented on this, my girlfriend said, "Where did the time go?" It seemed that we had lost track of some of the past couple of hours. Both of us were feeling rather exhausted for some reason and decided to just go home and call it a day, and we did. We told no one of the experience for fear they would think we were crazy, especially with my job. In my job, they would run you out of the place if you claimed to have seen what we saw.

Another reason I cannot go to anyone is because this is not my first sighting. Can you imagine what it would be like for me if I said it happened more than once?

In the early to mid-fifties, I lived in a suburb of a small city. I attended first through third grade there and afterward moved to a larger city.

In those days, we practiced air raid procedures in school and studied about protecting ourselves at home, during a nuclear attack. My stepfather was a member of the Civil Defense Unit there and was also assigned to the airport as what they called in those days, a "spotter." I recall him explaining to me then that his job was to watch for enemy aircraft, in the case of enemy attack. It seems so archaic by today's standards. Nonetheless, we all took it seriously.

One summer night, between 1952 and 1954, I can't recall exactly, the local air raid sirens activated and my father had to leave us and report to his assignment. My mother was extremely upset and my father left in a rush, wearing what I remember as a long heavy black fireman's coat and a white hard hat with an emblem on it, with the letters CD. I remember being afraid because they were afraid. I really wasn't sure what was happening.

After my father left, my mother directed me to assist her in turning the couch upside down in the living room and placing my two younger sisters under it. One sister was a year younger than I and the other was two years younger than I. After doing this, my mother directed me to help her fill water jugs and place them in the basement. I remember how heavy they felt to me.

When all the jugs were full and placed in the basement, my mother turned to me with a look of panic on her face and told me that she had forgotten to turn off the furnace fuel tank located outside the house. She asked me to go with her to do so. We went out the back door to the yard. I remember hearing the constant roaring of jet engines in the sky and in the direction of the airport. During those years, a squadron of F94 fighters were stationed at the airport. We used to go there routinely and watch them come and go, it was great.

As I looked toward the airport and the noise, I could see spotlights searching the sky, in a crisscross fashion. Periodically, the searchlights would stop in an area and seemed to concentrate there. I recall seeing their reflection off the clouds and seeing what appeared to be circular objects the size of a street light, caught in the searchlights. The brightness of the objects surpassed that of the searchlights. The objects (two) would remain stationary until the searchlights would lock on, then they would dart about very quickly and disappear completely, only to appear somewhere else in the sky. I had no concept of what I was seeing at the time. My mother became very afraid and proceeded to shut off the fuel tank quickly. Then quickly we both went inside.

It seemed like a long time went by before my father came home. When he did, in the predawn hours of the morning, my mother and I were still awake. I remember how calm he was, but I could sense a tenseness about him. He and my mother talked almost secretly about what happened and I was told to go in the living room and sit until they finished talking. I heard my mother ask him what "They" were and heard him tell her that he was not permitted to talk about it. She was very persistent about it. He then whispered something to her, I saw her face go pale and she fainted and fell to the floor. My father caught her on the way down, and gently lowered her the rest of the way down. It scared the hell out of me. I never found out what happened until many years later, in the sixties. My father, now deceased, never spoke again of the incident, but my mother told me, when my father was not at home one day, that the objects were "flying saucers." I later confirmed this in two books written about such matters. I believe one book was called *UFO Friend or Foe*, by a Major General, United States Marine Corps, Retired. The other book may have been *Project Blue Book*. This incident stuck with me the rest of my life, and sparked an unexplained, intense interest in these matters, mixed with an unexplained fear, which remains to this day.

During this period, in the fifties, I began sleepwalking,

almost routinely. My mother would tell me about episodes where I would actually leave the house and go to the yard. On one occasion, she stopped me from walking down the neighborhood street.

On most occasions, I would remember being awakened by her, after walking, but did not remember leaving my bed. I still remember these episodes to this day, and how the night felt so vast.

There were times, during the fifties, when my parents would place me and my two sisters at my grandmother's home. My grandmother would place me in an upstairs bedroom alone, and would place my two sisters together in another room. She was my stepfather's mother and she did not like me very much and did not hesitate to tell me. I was terrified to stay at her house, at night.

I recall having nightmares of someone coming into the dark room. At times, I remember waking up with a feeling of terror so strong that I could not move. Sometimes I would even wet the bed I was so scared. I could sense someone there but most times could not see anyone. On several occasions I would suddenly awaken from a sound sleep, frozen with fear. I managed to move my line of vision toward the foot of my bed, which was in the direction of the bedroom doorway. On at least one occasion, I felt as if I was still asleep but yet awake, like looking through a gray haze. I saw the silhouette of someone between the doorway and the foot of the bed, I would now guess to be the distance of about five feet away. I can remember being too afraid to even speak. I then must have just fallen back to sleep, because I never could remember anything beyond that.

My grandmother's row home was located on the corner of intersecting streets. On the corner of that intersection was a traffic signal, which stayed red unless you pressed a button on it to change it. During the day, my grandmother would allow us to go outside and play on the sidewalk next to her home. I would, on most occasions, refuse to go outside. I finally agreed to go out, if I could just sit on the steps to her home and not go near the traffic light. Strangely, I was terrified of the traffic signal. I could not even look at it without experiencing that feeling of fear, similar to that experienced in the bedroom at night. I have never understood my fear of this object. My grandmother pressed me about why I didn't want to be outside, and I finally told her about the traffic light. She more or less blew it off, but later used it as a means to discipline me when she felt I didn't behave well. She would threaten to take me to the light, and on one occasion, she did.

I don't know what I did that day, but it sure made her mad. She picked me up and carried me to the traffic light and held me up to it. I remember the green light lens was so close to me I could reach out and touch it. The red light was lit, but the unlit green light terrified me. I could not bear to look at it. She later reported my fear to my mother.

This fear stayed with me so long, in those days, that my mother finally took me to our family doctor. This fear of the traffic light is documented in his records. I became aware of the record in 1970, when I married my former wife. She was, at that time, a nurse for the same doctor and had seen the record. She later asked me about the notation about the traffic light. I downplayed it to her because I knew she would never understand. I didn't understand it myself, and still don't. To this day, I think about it whenever I see a traffic light, but it no longer scares me. Being scared of a traffic light would be a bad situation for a cop, don't you think?

ROTTEN APPLES FROM THE AIR FORCE

In Hamburg, as the child of a U.S. Army officer, I had a very vivid dream of being in a large plaza, in front of a large building. I was on my stomach, surrounded by thousands of young people who were also close to the pavement. There was a tremendous amount of machine-gun fire. I could see military tanks moving from the front and from behind. I felt the young man next to me holding me down. He was Oriental and kept saying things to me, but I couldn't understand him. I then realized that everyone around me was Oriental. I heard voices ringing out from bullhorns. The woman across from me screamed, then fell silent and I could see blood rushing from her head forming a large puddle. At this point I'm trying to figure out how to get away and notice as I scan the surroundings that military personnel are running barbed wire in different perimeters. When I saw the slaughter of Tiananmen Square years later on CNN, I began to cry uncontrollably.

In college, my thoughts were on women and composing music. I came home around 2:00 P.M. from college with severe pains in my groin and bowel region. I sat down in my room, leaning against the wall when I heard a massive sonic boom. Immediately, a star burst occurred in front of my eyes and my vision was changed to a perfect rectangle, which was shrinking away while the star burst increased, as if I was traveling at the speed of light. I remember them turning into billions of needles of multicolored lights, when suddenly the rectangle returned, containing my front view of my room in full vision. I realized I was completely paralyzed and staring at my closed door, when the door suddenly opened. I was in a state of shock, just vegetating, when I was suddenly flipped forward, then 180 degrees around, so that my body was hovering a foot above the floor with my face to the ground. All this happened in a machinelike manner. I then began floating out the door, through the hallway. I passed my parents' room, where I saw my dad napping, and into the kitchen where I was seated in the dining-room chair. In front of me was a small glass and an egg. As if I knew what to do,

I flexed all the muscles in both my hands and made both objects float in the air. I then willed the egg to spin slowly and manipulated it into the glass. I sat back and enjoyed my effort. I then snapped into awareness and found myself in the same place in my room where I'd begun. At first I didn't believe it to be a dream. When I initially came home, my father wasn't there. So I dashed to his room, finding no one. Strangely enough he arrived ten minutes later and took a nap in exactly the same position I'd seen him in my dream. All my pains had disappeared.

After a month, for no apparent reasons, the strangest type of dream began to occur at night; first once a week, then two to three times a week. I would be laying in bed close to or past midnight, when I would hear cracking noises traveling uniformly from my bedroom wall across the roof. It sounded like the walls were having conversations. I would hear a small explosion, then a massive sonic boom at which point I would immediately go into paralysis. Then I would slowly begin to float toward my ceiling, unable to move a limb. The episodes would seem to last no more than five to ten minutes. I soon began to regain a small amount of movement. After reading extensively on the subject of astral projection, I was convinced that I could learn to control it, and maybe even travel further than the confines of my room. No matter what I tried, they would happen when they wanted to and not when I did. During the summer, I pulled out my Xerox copies of journal clippings on telekinesis. I tried a simple experiment. I placed my keys on the center of the living room table and began to concentrate on them, wishing them to disappear. I sat there for almost an hour focusing, sweating to the point of nausea. I stopped, gave up on the whole ordeal and began washing the dishes, when my uncle came by. He let himself in (he had a copy of our keys) and went to the temporary room he used while we were on vacations. I noticed the keys were gone from the living room table. Next began a three-hour search for my keys, in which my uncle helped and his friend who arrived soon after. They were forever lost. Again I dismissed the whole thing.

The following August I experienced one of my last lucid dreams of floating. It was the same scenario, except this time I floated out the door, through the hallway and into my parents' room. I was floating with my back toward the ceiling, when the room suddenly filled up with a fantastic, blue fluidlike light. I saw my mother open her eyes, and upon seeing me, her features changed to a face in terror. I could not make a sound. Though I saw her mouth move and her arm stretching upward and her hair falling toward me, as though static electricity was in the air, I could do nothing except stare. In a quick animated motion she fell back to sleep and the room turned dark. After this night I decided not to dream like this anymore. When I heard the first bang, I would get up. They stopped completely until the following November. The typical scenario unfolded, except this time I was fighting and pleading for "them" to let me down. This time I could feel a definite force wrestling with me.

My proof that there was more to this than just dreams came right away. While struggling I managed to turn my body to look down. For the first time I could see that my body was not in my bed. I panicked, realizing that this was not astral projection and reached to hold on to the top of my antique china cabinet. I could only grab my T-shirt, which I put on the top every night, and flung it toward the bed, immediately reaching again for the cabinet. I was pleading the whole time with an entity that seemed to be invisible. The "dream" stopped abruptly. I could no longer sleep that night. When daylight broke, I realized that my T-shirt was not lying in its right place. I searched the bed. Nothing. I then recounted every detail of the dream. The force with which I'd flung it would have placed it in the corner of the bed and the two adjoining walls. I stood there for a good five minutes, scared to even consider moving the bed. Well, I did, and there it was.

My wife was seven months pregnant and I was on top of the world. It was June 1995 and once again I floated, paralyzed after I heard a loud bang. Only this time, I was fully conscious. I decided immediately to fight the fear and go with it, and to my surprise I finally went somewhere. After being lifted out of my bed and onto my feet, I floated toward the right wall. Realizing that I was going to be crushed against the wall, I blinked and found myself slowly going through the wall. The wall was translucent, bluish in color, allowing me to literally see the interior structure. During this process I heard a continuous sound that reminded me of a hand rubbing against a balloon. After I passed through, I was in a different place that was filled with staticlike energy. I then passed through several walls of extremely bright white light, when I suddenly passed into daylight. My wife was next to me. Together we followed an elusive guide who seemed human. We were joined by two or three other humans and entered what seemed to be a military installation. There were several barracks. We passed through each door observing the busy buzzing of scientists in white smocks and uniformed personnel. The strange thing was that I was floating and not being noticed by anyone. I remember a two-star Air Force general whizzing by me. I looked at what some scientists were doing, and saw them mounting an intricate part into what seemed to be a guidance system for a small fat rocket or warhead. I don't remember anything else beyond that.

I awoke in the morning feeling that mankind had failed to stop the few rotten apples who have spoiled the lot.

A STUDENT

It is really very difficult to write to you. Many times, my compulsions to communicate with you occurred after reading chapters in your books. Now my heart yearns to tell you that our spiritual journeys of awakening seem to pull me inside out, rather like folding into myself. Like you, I am deathly afraid and extremely excited at the same time, of some of my clear memories of the visitors. Admittedly, I am ecstatic, because I believe myself to have been one of the children you showed kindness and tolerance to over forty years ago, in some sort of school setting during an abduction episode. My younger sister and I pleaded to stay with you; we did not want to return home.

You were wearing a floor-length tunic with folds, kind of gray-beige. Your hair was wavy to curly, and reddish brown.

My mother found it curious that, months before I was to start first grade (I never attended kindergarten), I appeared to have already attended school. I recalled that my teacher had been a tall young man. Of course, over forty years ago in Albuquerque, there was no such thing as a male first grade teacher in any Catholic school run by the Sisters of Charity!

I'm so glad that, through your books, you are still teaching by writing what you have gone through. And that what some of us are going through is not isolated.

I must confess that I have an unnatural fear of the dark, because I can usually see the visitors clearly at night. Most importantly, my very being is so happy to remember you. Thank God my husband is my anchor back to reality. Like your wife, he understands.

A Teacher

I have had plenty of UFO experiences. My memories are all pleasant. I know well the lady on the cover of *Communion*, and I know her husband. My memories go back to childhood.

One of my first memories is of being in a classroom with other children. We were a select few. I was very happy to see my teacher, the lady on your cover. I loved her teaching. I was about the age of nine.

My next very clear memory comes at age eleven. I remember being in her office. She has an office down at the end of a hallway. It was like any ordinary office. Her husband was also there. She had a radio on her desk that was playing my favorite kind of music. The three of us were dancing to this music. They were actually teaching me how to dance! They were so encouraging! I was looking up into their eyes, and I just adored them. They looked at me like proud parents. I was wearing summer clothes, and I have no idea how I'd gotten there.

My next memories are more vague. These things occurred between the ages of twelve and twenty. I had many evenings of feeling that I was being watched, observed, and examined. I got so used to it that it didn't bother me. Occasionally, I'd wake up with little gray people around me. I never associated them with UFOs. As soon as I'd open my eyes, they'd all run away, right through the walls!

I was about fifteen when I saw what I thought was a skeleton materialize down at the end of my bed. This was one of the few times I was frightened. I now know who this was.

On a couple of occasions, I'd wake up paralyzed. I couldn't even open my eyes. I'd sense people around my bed, but I didn't seem to mind. I'd start to see the most beautiful geometric patterns. After this test, they'd tell me I had passed.

I remember having great explosions of light in my head, accompanied by a crackling sound; it was sort of like lightning. I also had tremendous psychic ability as a teenager.

I was eager to go to bed at night, because I'd slip into a world of the most amazing geometric symbols and designs.

They'd run like a ticker tape, black symbols on a reddish background. I felt that I was learning at night.

When I was fourteen, I saw and had telepathic contact with a UFO. I had a witness to this. My girlfriend and I were out walking one evening when I began to have the feeling of being watched. She didn't have the same feeling. A few seconds later I saw what was watching me. I could feel it talking to me and testing me. The left part of my brain was being probed. I screamed and pointed, and my friend looked up just before it disappeared. I got a very good look at it.

When I was twenty-three I woke up one night to find a little gray man on the other side of my room. He looked about four feet tall and had very large orange cat eyes. I later learned that this was my "guardian."

I was about the same age when I went through a big change; I became a Christian. This was a decision that would have a tremendous impact on me. I was happy with my decision, but suddenly everything I did seemed to go wrong. At night I'd find myself in a "hallway" like I usually did, but now my presence brought shrieks of terror from my "old friends." I wasn't welcome anymore; it seemed like I was a traitor. However, my "friends" reserved the right to torment me. I had to learn to fight, or they'd torment me to death, so that's what I did. I learned what made them jump and I actually came out of that world victorious. This took me years to do, and I had to do it myself. There was no one I could turn to who'd believe something crazy like that.

About eight months ago something strange began to happen. I couldn't get the picture of the being on the *Communion* cover out of my head. I'd long since put my unique past behind me; I'd even forgotten what my old friends looked like. I couldn't figure out why I kept seeing that darned image in my head; I'd only seen the book in stores. This went on for two months before I finally broke down and bought the book, just for the cover. After reading it, I went to the library and checked out anything I could find on the subject. It was at this time that some of my memories surfaced. It finally occurred to me that you were living in a world that I was once a part of, only you and so many others I was reading about weren't happy, as I'd been. I also remembered what I'd had to go through to get out of that world. What you were going through just didn't seem right. One night I decided to pray for you.

I really wasn't expecting anything to come out of this. Was I surprised! On night three, my prayer was heard. I'd no sooner closed my eyes when everything went very black. I then found myself in a well-lit hallway facing an old friend. God wasn't the only one who'd heard my prayer! She was standing in the doorway to her office. She studied me for a moment. I remember saying, "Lord, look!" With that statement, she ducked behind the wall. She didn't want to be revealed.

Again, everything went black and I was back in bed. About twenty seconds later I saw a small gray box in front of me. There was a hand over it, also gray. It was a clawed hand, and it held a silver wand. The top of the box was tapped, once. This created a great wave of knowledge that went through me. Physically, it felt like a slow-moving ocean wave. In that wave was the knowledge of myself. I also knew that I'd seen the gray box many times before. It was like an old, familiar friend. So was the being that had tapped it. I was now fully aware of the other world and life that I'd been in. I had had no idea before this about the gray box or the other life. I was shocked, and didn't sleep until very late.

"She" had shown me her power and authority. But now I also knew a greater authority.

The next few weeks were turbulent, and I saw a lot of my old friends. Some of them even came to inspect me and see who had uncovered their world.

I had another unique memory surface. This memory came in the form of a vivid dream. It involved being brought to a college campus. It was a very familiar road to me. One night I was brought down that road to meet you. "They" wanted me to get you to talk. You stayed in a small room and wouldn't talk to me; you seemed very sad and upset. Even though you wouldn't talk to me, we did study together that night. I seemed to be about fourteen. I remember not liking the way they treated you, but what they did to you to upset me I don't remember. When I awoke, I felt as if you had been looking at me intently.

In regard to the symbols that I once lived with, something has tried to surface. I started to get what appeared to be an alphabet, a logic system, and a time system.

I know that I was manipulated into buying a book that I never had any intention of buying. I was also manipulated into praying for you.

CHAPTER FIVE

Journeys

I saw Eternity the other night,
Like a great ring of pure and endless light,
All calm, as it was bright;
And round beneath it, time in hours, days, years,
Driven by the spheres. . .

"The World"
Henry Vaughan

Journeys: The Voices of the Old Hands

One particular type of letter we get describes journeys with the visitors. These letters have very similar characteristics in that their authors have usually had many years of conscious contact experience. These are generally people who have been interacting comfortably with the visitors for most of their lives, so these may be stories, in a sense, from the future of mankind. Could there come a time when many more of us will live as these people do, with commonplace reference and access to worlds beyond our own?

Issues of fear and consequent traumatic amnesia generally don't enter into these narratives.

How real are the stories? As incredible as they are, there is no more reason to dismiss them than any of the others. One of the great problems with the visitor experience is that anything that doesn't fit expected patterns is generally ignored. Insofar as we know so little about the origin of the experience, much less the nature and motives (if any) of the visitors, this seems foolish.

Far better to remain open about everything, bearing in mind that the most fundamental issues are still in question. That way, there is no need to pick and choose among narratives.

Using this much more reasonable approach, we can open our minds to these wonderful stories of journeys to truly exotic and faraway places.

Many of these stories are characterized by elements that reflect the old fairy lore. For example, the journeys may seem to take long periods of time, but when the witness returns only a few minutes will have passed. In one case, two witnesses enter upon a spring day in the middle of winter, an experience familiar to the fairy tradition, and one which I had myself when I was nine. This experience, related in *The Secret School,* involved stepping out on a porch in the middle of the night to find that it was suddenly midday—but not on another planet. Apparently, I had gone to the future.

In my own memory, there is a journey to another world, but it is so dreamlike that I have never included it among my factual experiences. I have alluded to it from time to time, and used it as the basis of the fictional journey to another planet that occurs in my novel *Majestic.*

I have also seen a fragment of videotape that is said, by the man who possesses it, to have been made on another planet. It is a truly astonishing piece of tape that depicts a sunrise on a rocky shoreline beneath a double star. It is so complex and so vividly realistic that it is hard to believe that the man who originally made it had created a fake, especially because of the very substantial special effects resources that would have had to be involved.

Like many of the witnesses whose narratives appear in this section, this man had a long-term and comfortable relationship with the visitors.

Why would there be certain people like this, when the rest of us are having such a hard time? Perhaps the best thing to do is to read their reports, and reflect on how much fun they are having and what a remarkable sense of adventure seems to characterize their lives.

THREE DAYS TO ANOTHER WORLD

I want to say that I admire your courage and to thank you for writing *Communion*, *Transformation*, and *Breakthrough*. I include Mrs. Strieber in this because I believe she is just as worthy to be recognized as Mr. Strieber because of the experiences you both are going through. I know that you get a lot of mail from people looking for the "all-knowing answers" that they think only you have. I don't want to be one of those people. I would like to tell you briefly about my own experience with "the visitors." I have only been taken by them for about three months, in fact, it happened the night after I finished your last book, *Breakthrough*. They come for me at least twice a month. I don't know why they want me so much, I think that they need me to help another person deal with the takings. The other person is my best friend Tom. They've been taking him for twenty-eight years. He feels better now that there is someone who understands him and what he is going through. I feel kind of like Mrs. Strieber, who is your comfort through this. I am glad I can be helpful to Tom.

Last month, Tom and I were taken aboard the ship—it was a small ship, not like the one I was on before. We were told that there was work to be done and that we were needed to assist them. (By them, I mean the "grays.") We were gone for three days.

On this ship, there was Tom, one half-breed, five grays, one blue, and myself. We went to a beautiful planet (I think it was another planet) that was very hot. It was hard to breathe. I remember seeing a flower and wanting to pick it up. They said not to because it could harm me. We were on what seemed to be a mesa. It was filled with vegetation, but no insects and no animals were present. Tom remembers picking a fruit off a nearby tree that resembled a pomegranate. The grays said we could eat the fruit, it wouldn't hurt us. We were told to scatter the seeds of the fruit when we were done. I'm sorry that I don't remember the task that we had to perform there, but I do remember being told that we (I mean all of us that were on the ship) did an excellent job. We went back to the ship and started for home. On the way home, I remember asking many questions.

The blue came to me and said: "Would you like me to answer your questions now?"

Of course I said yes. One of the grays took us to a terminal and the blue put his hand onto a control panel. I was then allowed to look up anything I wanted! I looked up different beings, planets, plants. It was excellent! Just before arriving home, the grays came to Tom and me and told us that we have achieved a "higher rank" because of the good job we did on the planet. We had no clue as to what this meant. Next thing I knew, I was home. According to the clock, we were only gone for an hour of our time! How could that be?

Enough of bothering you with my story—I know that you and your wife are very busy people. I'm sorry I took up so much of your time. My one question is: Why did the visitors give Tom and me what seemed like carte blanche on their ship? Tom and I even gave out some orders! I am very confused. They have never hurt me, they have never scared me.

YOUR WORLD IS ENDING. WE CANNOT INTERVENE.

My younger sister, whom I'll call Barbara, was involved in a childhood experience that has always stood out in my mind. As children, we often played together on the steep wooded bank behind our house. One cold winter day we climbed into the woods and found a small clearing. In the blink of an eye, it became a warm spring afternoon! Birds were singing, there was green where there'd been snow, and it was warm enough to remove our jackets. We sat on a log and enjoyed a picnic. We remained for some time then donned our jackets. Outside the clearing it was cold winter again. There was something more, but I can't remember it.

Neither of us ever spoke of this again until we were adults. By then I imagined it to be a dream, and I actually mentioned it in the context of a dream to Barbara. She looked at me point blank and said, "That really happened; it was not a dream. I remember it. Wasn't it weird?"

I also have a recurrent dream in which I return again and again to an idyllic spot near the Hudson River. There is a field of flowers, some old buildings, and a deep feeling of home. I don't know if this place exists, but I have a strong conviction that if it does it's in another dimension of time and space. I know I've been there, and in my dreams I'm usually running before I arrive there.

During the early seventies, I lived in upstate New York. While out on a summer evening ride with my husband at that time, I called attention to a large metallic object, clearly hovering near the Shawangunk Mountains. [Note: Where Whitley had his cabin.] It was still daylight, and we were both amazed at such a large object in plain view.

It was also in the early seventies that I began to experience odd OBEs. Often, during one of these, a dark opening or hole would appear. I'd then enter it in my nonphysical state and travel extremely quickly to other places in space/time. My memories of these excursions are now vague and somewhat confusing, but at that time I had very good recall. I was, and am, definitely of the opinion that I was being taught. In one experience I entered the body of a man living in another space, time, or planet. It was very green, and it was an agricultural-based society. I lived this person's life as a husband/father/community elder for what seemed to be about two months of our time. During this time, he coexisted in the body with me, yet I felt everything he felt and made his decisions, etc. All of the machinery was steam powered, and it seemed to be an idyllic society, but there were undertones of class discrimination. Imagine my amazement, after experiencing all this, when, upon returning home, I found I'd been gone about twenty minutes!

I spoke with beings of other dimensions and planets; at times I seemed to have a guide. There was some sort of interplanetary or interdimensional tribunal. I only partially recall these things now. I remember vividly the black hole appearing for me to travel through, and the vibratory sensations in and around my body as I separated. I also remember intense fear, which I've never overcome, when even now I experience an OBE episode coming on.

At one time, some beings with bald heads were staring at me and one was saying, "She's not ready yet." During this period of intense activity, I also dreamed vividly of sinister people in black sedans.

I feel that the visitors contacted me through OBEs, and now more than ever, in the dream state. I believe that they're contacting many people simultaneously, and in many different ways. Each of our souls/spirits is different, and we are reached through whatever channel is most conducive to our needs.

In the early eighties, I began a search for the meaning of the multidimensional adventures of the early seventies. My questioning was often answered with blank stares or mystical rhetoric that seemed to have no relationship to my personal experiences. During this period, I again witnessed UFOs. I was living in the Hudson River area at the time. Looking up at the summer sky, I saw seven small disks flying overhead. I felt very excited, yet peaceful. The sighting seemed to have a reassuring effect on me.

A few weeks later my younger sister called, very frightened.

She'd been visiting me, and on her way home that evening via a back road, she spotted a large and well-lit UFO. It approached her slowly, then hovered for some time before darting off.

I have experienced voice phenomena, also. I've heard a male voice, very authoritative, and a gentle woman's voice. The male voice tends to warn me of physical imbalances or impending illness, and the woman's comforts me in times of emotional distress.

When I saw the cover of *Communion* I felt compelled to buy it. When I began to read it, I felt nauseated, burst into tears, was shaking, and was elated. Most books don't elicit this reaction in me as I read the first few chapters.

I have suffered my own growth in strange ways. I remember waking often between 3:15 and 3:30 A.M. in a panic and with a fear of my own backyard in the night. I remember dreaming of trips or visits from UFOs, and waking sore and tired in the morning.

I have a friend here, an artist, who told me that I appeared at the foot of her bed in the middle of the night. Sparks of light drifted from my fingertips to hers. Some time later, she witnessed the same grouping of disks in the summer sky as I did several years ago. With them came the same feeling of peaceful reassurance.

I believe I'm still being contacted and taught. In about 1986, I dreamed that a large bubble landed next door to my father's house. Out stepped a small being. I expressed my amazement at his childlike appearance. "No," he said in my mind, "On my planet children are the adults. Come, I have something to show you." I stepped into the bubble and we exited onto another planet. He opened a window, and through it I could see planet Earth, very close. I expressed wonder that no one had ever seen this planet, as it was so close to the earth. "No," he replied, "we exist in a parallel reality. Look—" The polar ice caps on the earth were melting, causing massive flooding and imbalance. He said, "This will affect the tilt of the earth on its axis. We cannot intervene." Then he brought me home.

This is as I remember it, though it's somewhat distorted

and there was much more. I sensed great danger for our planet, and there was something about ozone. I've had many "alien dreams" since then, and though I still have an occasional OBE, I don't recall any more contact in this way.

Last year I attended a party with some friends. While sitting on the deck that overlooked the valley, I recounted a dream I'd had a year earlier. I dreamed that I sat on that deck (which I'd never seen then), and witnessed a display of UFOs. After much laughter and a few drinks I went home. Ten minutes later the phone rang. The remainder of the party was on the deck viewing a fascinating group of UFOs. Because of trees obscuring my property, I couldn't see them, but they did make the evening news. Someone had videotaped them.

I love these beautiful, fragile human beings around me, and yet I know that a transformation is taking place. We may be a doomed species. But we will not end; we will only emerge as a butterfly from a cocoon, or the Phoenix from the ashes.

"TERRIFIED BEYOND DESCRIPTION, BLESSED BEYOND WORDS"

Your description of the visitors "taking you along" for the visit to the little girl and her mother [in *Breakthrough]* is astoundingly similar to one of my own experiences. Not only have I never heard of anyone other than myself being "taken along," but there are a number of similarities which reopen the question of the meaning of the experience for me.

I was taken along on a house visit to see a little girl who was described as ill. I watched while the visitors did some adjustment to the girl.

My first group of abduction experiences happened in 1976. The incident I'm about to relate happened around 1989. I have a terrible memory for dates, but it's dated somewhere in my journals, and also confirmed by an incident which followed, which involved others. I was taken by a group of four or five human-looking people. I was told, and compelled, not to look at any of them, except for the main contact person. For this reason, I cannot be sure that all of them were actually human-looking.

It was probably 3:00 A.M. when they came into my apartment in San Francisco and woke me. I felt oddly happy, as if seeing old acquaintances. I called the contact person the "blond guy," because he was around five feet seven inches, with dark sandy blond hair and golden eyes. He led me to the window, and had me look out over the city. The others stood behind a lot, and seemed to be "watching my thoughts." Communications were telepathic. The blond guy asked me, as I looked out at the city, how I felt about my work, life, and relationships in human society. He asked me to think and feel this as clearly and truthfully as I could. They seemed to watch my feelings. He then touched the base of the back of my head and seemed to search (with his fingers?) for three spots in a triangle shape. I could feel him finding them, and then he pushed, or exerted pressure of some kind. I immediately felt a wave of relaxation, then I sort of blacked out.

I then had the sense, which I've had during previous abductions, of traveling or moving through darkness. I seemed to be conscious, but without vision. When next I could see, we were in a building which looked like a bunker. (After searching seven years I found this place, which I had never seen before except during the abduction. It was the exact same place.) I think they did some procedure with the implant I had received in 1976. I have the feeling that they removed it. I was very afraid of the pain. To overcome this, the blond guy told me to look into his eyes and only to focus on him. I did, and he radiated such a sense of brotherhood and compassion that I felt I'd do anything for him. He seemed to see through my soul. Then they performed whatever the procedure was.

They then walked me to a dark window. This didn't seem like a normal window. The next thing I knew, we were traveling. I noticed none of the "car" details, as you did, but I was conscious and aware of the presence of the group of visitors who had been attending me since my apartment. They told me we were going to see a little girl who was in some way ill, not necessarily physically. They gave me a very brief sort of psychic profile of each family member. The mother was in some way important to them, and the father they did not like. I received an image of him as aggressive, dominating, arrogant, and somewhat sadistic. They told me it was absolutely imperative that I do nothing other than exactly what they told me to do. I realized I was being taken on a home visit, abduction, or whatever I should call it. I would be in the position of observing people in the position I had been in myself. We arrived, but I don't know how, outside of a suburban house that could have been anywhere, but I felt it was the Midwest. It did not seem like California. It was so similar to what you described that I nearly felt faint when I read your description. The time of night seemed to be around 3:00 or 4:00 A.M. I felt the cold night air, the stillness, and everything was completely real.

We went to a sliding glass door in the back of the house, and somehow passed through it. Maybe they opened it, but that's not in my memory. We walked through the dark hallway. I seemed to share the visitor's knowledge of the floor plan telepathically. I sensed where the little girl's room was. We went to the parents' room. A couple of visitors went in who seemed to want to check on the mother. They "said" the father could be dangerous if awakened. I could feel their dislike of the constant, unhealthy influence he exerted on the mother and little girl, but understood this could not be helped.

I stood away from the bed, hoping the father wouldn't wake up. There were either two doors to the girl's room, one from the hall and one adjoining the parents, or what I thought was a door was a closet and the visitors went through the wall of it. One way or another we went into the girl's room. As she slept, one of the visitors, whom I recall as being a woman, went to her bed and somehow prepared to adjust her energy or something. This would not cure what was wrong, but it would help. It seemed this problem had lately been bad. They had visited her to help her before, and would again. I stood looking at her sleeping, and felt her struggle with the father's bad influence, which seemed to play a large part in whatever was making her unhealthy. I could feel the mother's deep worry, which she had to suppress around her husband. They led me out through the house, and the woman stayed behind in the girl's room, I thought, to watch over her.

In the darkened living room, I noticed the furnishings. I felt the family's life there. I knew the man would likely shoot us if we were normal intruders, but I knew the visitors would make such an action impossible. I was safe, but how strange to be in a stranger's house, knowing what I knew. They led me outside again. Some of them were still in the house doing things with the mother and little girl.

They directed my attention to the house next door. They showed me a telepathic image of a woman in her fifties who lived there. I saw an image of her looking out of her window on some earlier occasion. She had seen the visitors with the little girl, outside during a night abduction. She had suppressed this memory, and had transferred it to a persistent and strong feeling of anger and resentment toward the girl's parents, whom she considered to be terribly negligent. She feared this would result in the little girl being kidnapped. I think I was shown this to give me a sense of the many emotional dynamics and complications of the visitors' activities. I had the sense that this was, in a way, training for work I might one day participate in, perhaps beyond this lifetime, or in sleep. My memory was of standing outside the house of the family, and then we just seemed to be back at the bunkers.

They took me into another room where there seemed to be a committee of higher ranking, or elder, visitors. They would oversee the next aspect. In brief, I was led to a window where I remember putting my hands on the slanted concrete sill. They telepathically asked questions. As before, they wanted me to feel the answers as deeply and clearly as possible. They stood behind me, seeming to look into the back of my head. I didn't look at them. As I looked out the window I saw the Golden Gate Bridge. They then told me to go to the other window, out of which I saw Alcatraz Island. I then saw, as through a telescope, two men swimming in the bay, escaping the prison at night. One man began to drown and the other turned back briefly. Should or would he try to save his companion and likely lose his own life, or should/would he do the painful but reasonable thing and save himself? This was the question they seemed to want to know my feelings about. The scene, and the horror of that swimmer's dilemma, has haunted me ever since. It occurred to me later that they were testing my sense and level of compassion. (Seven years later, when I found those bunkers, I went to the windows and placed my hands on the concrete sill as I had that night.) I don't recall leaving the bunkers, but after the Alcatraz vision there was some conference with the committee and the team in charge of me. I was, I think, placed on hold, in a semiconscious state.

In my next memory, we were all standing in my bedroom again. I had the sense that at least two hours had passed since the experience started. They told me that, when they left, I'd get into bed and go to sleep. They left by the front door and went down the hall stairway, just like guests leaving. I accompanied them to the stairs and said good-bye. I felt sad; I had a sense that I probably wouldn't see them again for years, if ever. As if to offer consolation they said, "Sometimes, in the city, we are nearby." I walked back into my apartment, closed and locked the door, lay down on my bed as they had told me, and closed my eyes. In what I think was a second or two later I opened my eyes again and looked around. The room was just as it had been a few seconds before, when I'd gotten into bed. I got up and walked to the places where we had stood. I went out the front door into the hall and looked at the stairway where, it felt to me, the visitors had exited only a few minutes before. I don't believe I was asleep at all, and I think the events I've described transpired exactly as I experienced them.

A week, or perhaps only a few days, later I went to an odd little UFO lecture held in a hotel, sponsored by the Junior Chamber of Commerce. An acquaintance had invited me out of the blue. I was purposely avoiding UFO matters at the time.

Because the "blond guy" had told me to look into his eyes and focus only on him in order to calm me for a procedure, his face was burned into my memory, especially his odd golden-colored eyes that were sort of like the color of a monarch butterfly's wings, or a lion's eyes. A minute before the lecture began, I spotted him in the crowd!

All I could think of was confronting him, so I don't remember the lecture at all. During a Q&A time, he asked about the status of something I'd never heard of. The name sounded odd, like some code-named project, not something he could discuss. The blond guy then asked a question about some specific people, naming a couple of names. The lecturer seemed at a loss for words and looked at him and said, "Who are you?" The blond guy didn't respond. The lecturer said, "Because you're asking questions about things you couldn't know about. I know all of the people involved who know about this, and you're not one of them. Where did you get this information?" The blond guy did not say anything. The lecturer said, with a nervous laugh, "Are you a spy?" The guy said, "You might say that." The lecturer said emphatically, "Will you stay after I finish here? I really need to talk with you. Will you please stay?" The guy didn't say anything. The lecturer seemed spooked and distracted, and tried to finish up the Q&As. The people in the audience seemed sort of titillated by the exchange between the lecturer and the guy. I can't describe this mood. Can you imagine my state, given the fact that I recognized him as the main person in my "abduction," and then had seen him visibly spook the UFO expert?

As soon as the Q&As ended, I went as quickly as I could to the blond guy and said, "Can we please talk? I have something very strange I want to discuss with you." He said, "I don't think so." I didn't feel I could just say, "I recognize you—etc.", although now I wish I had. I said, "Please, I really need to talk with you!" He said, "I don't think that would do much good." I was thrown off balance by his odd answer and said, "I really think I have something to tell you that would be of interest to you." He said, "I doubt it." I flustered and said, "I really think it would be—" He said, "My range of interest is very narrow." I asked, "What is your interest?" and he said, "My interest is implants." He paused before the last word, which hit me like a punch; it felt like he had delivered it that way. Without another word, he turned and walked into the crowd. I was speechless, literally dumbstruck. I had looked into his eyes during my abduction, and felt such deep compassion and caring that "for him" I submitted to the implant procedure. Now his last words had left me speechless. After a few seconds, I realized I had to confront him, or I'd lose the chance forever. I followed him, pushing my way through the crowd. I lost sight of him at the main doorway of the meeting room, only about seven feet behind him. When I got to the door, I asked the guard if he'd noticed a blond guy leave and he said, "He just went out a second ago." I went into the huge main hotel lobby. The nearest exit was seventy feet away, the nearest elevators forty feet away. There were no other nearby doors at all. Even if he'd broken into a run after exiting the meeting room, I couldn't imagine how he'd gotten out of view. I looked into the bathroom, and searched the inside crowd again, but he was gone.

It didn't occur to me to ask the lecturer what had upset him about the blond guy's questions. I doubt he would have told me, anyway. I should have told him my story, but I was embarrassed. Not all UFO investigators were, in those days, open to so-called abductees, as you may recall.

Why would we, in both our cases, be taken along to see a mother and little girl who needed a health adjustment? Why did we both receive, almost, a sort of proof within a few days of the experience?

I feel that a good deal of what I've seen the visitors do has been a kind of theater, constructed for the effect it had on me, rather than an expedient of their needs. If this is all or partly true, what symbolic meaning would there be in our both having been taken to see a little girl, in my case, said to be unwell, and in your case the girl's illness might have been instigated by the treatment? Could the scenario have meaning beyond the actual events? Perhaps to more people than you and me? What then? My own experience instructed me, in my own capacity, about compassion and empathy. I felt this very deeply, and recalled it for years.

In my 1976 experiences I was abducted repeatedly by tall and short alien beings. I was confronted by a being who must have weighed 800 pounds and who was ten feet tall, heard an incredibly deep voice speak my name, had black helicopters appear to look into my windows. I saw a man who looked like a seven feet tall corpse in a 1940s suit, with a three foot tall friend in white hooded overalls, goggles, and a bizarre red wig. They were pretending to mow the lawn of an abandoned house. In short, I've seen some pretty damned weird things. I was also given mental exercises, which I thought had made me psychic. I certainly could hear people's thoughts. I heard trees make a sound at sunset which I call "singing." Most of all, I felt compassion that I'd never known before. Most of this abruptly ended on the day the intense period of my experiences stopped. I believe they were inducing "assisted telepathy," and had shown me the exercises to prove that these abilities were rationally arrived at, and real. They cautioned me against faith in anything I could not "take apart and put back together again" myself. I have been terrified beyond description, and have felt blessed beyond words.

At present, I'm embarrassed to say that I'm being plagued by nightly visitations from swarms of moving shapes, which I very hesitantly have to call ghosts, or other-dimensional beings. They seem directly related to the visitors. I have a constant sense of impending disaster: earthquakes, tidal waves, riots? Perhaps this is just the Post Traumatic Syndrome Disorder (PTSD) with which I've been diagnosed. I also have an intense fixation with sacred geometry and the laws of reality that I find embedded in number sequences. These fixations, unbelievably deep and unexplainable, emanate from my visitor contacts.

I continue in my work as the art director of an internationally known theater, and maintain a sense of composure in my quite stressful work. No one ever knows that I go home to an apartment swarming with other-dimensional beings, feeling the world is about to fall apart. Just a very normal life.

FLIGHT TO THE PAST

I have gotten through almost half of your latest book, *Breakthrough*, and a couple of things have caught me off guard. For example, when you stated that after you decided that you wanted to be with these beings when they visited another person, you stated that you were "taken" to what looked like a vehicle on your front porch. I'm sorry, I have to stop typing for a bit, I can't believe how much I'm shaking.

I have a memory of when I was eight years old. I remember waking up on my grandma's side porch. My grandma lives on an Indian Reservation in South Dakota. It is not a very populated area. Anyway, I woke up surprised to find myself sitting outside of the house. I had no idea what time it was, yet I knew it to be late. What scared me was the fact that my family was not aware that I was no longer in the house.

My grandma lives in a two-bedroom house. At the time she had one bedroom and my family, six people, had the other. The five children would sleep on the floor while my mom slept on the bed. It was almost impossible for us to leave the room without stepping on somebody. Which explains why I was so surprised to find myself outside without anybody noticing I was gone.

As I looked up, I realized that someone was coming around the house. I watched this person, not feeling fear. I remember looking at this person, and not being a bit surprised about how he looked. He didn't look like one of the people pictured on your book, *Communion*, but I knew he was different. He gave me what looked like a smile and I remember that I knew his name was Curry. Later on, I discovered that curry was actually a sort of spice from India.

I remember that he wanted me to go with him. My job at this time was to calm a person who was in a vehicle. The way I described it to my mom was, it was a black car, and when you opened the big door, you could see that there was only one door. As I walked up to the car, I could see the person huddled in the front seat. He was sitting up against the far window. He must have been at least twenty years old. He might have been older or younger than that, but when you're eight, everybody over ten years is old. What shocked me is that when he looked at me, he didn't really look all that much different from Curry. He also had huge eyes that weren't black, but he had more hair.

I climbed, literally, into the car. It appeared to be high off the ground. I remember reaching out and taking this person's hand, rubbing the top of it and saying, "Don't worry, they won't hurt you. They are really nice. You'll be all right," (or something to that effect).

After the door was closed I remember that we drove (flew) to a crossroads out in the middle of nowhere. When I see the crossroads now, I try to look for a place where a ship or shuttle may have landed. I held this person's hand as we walked into the ship (or shuttle). I remember heading for a "window" so I could peek out while we lifted off the ground. I don't remember looking around the ship. But I do remember seeing the ground get further and further away. One thing I remember, also, was when I was dropped off, I looked at Curry once again and this time he was wearing a hood. He looked kind of funny because all I could see was his eyes. I'm starting to think that maybe the scared person was actually one of these visitors whose job it was to keep me from getting too curious by distracting me from the other people in the car.

I really appreciate my mom because when I told her the story, she made me feel really special. She did not shun or try to find logical explanations to my story. I remember she cried after I told her, and she said that she was glad that I had made new friends. Later, when I was a teenager, I remember she wanted to know if, on one of these excursions, she could come along. Or at least meet them. I told her I would ask, but it would appear that these visits are only to see me. I don't recall that they ever responded in the affirmative.

One of the other experiences I had also dealt with time travel, I think. I also remember waiting to go with them to see what they did on these visits. When my family lived in Boulder, I remember waking up in the backyard and looking at this vehicle, again. I got into the backseat because that's where I remember being assigned to sit. I have the feeling that this wasn't the first time I went with them, but it was the first time I consciously remember asking if I could go with them.

We "flew" over this weird-looking landscape, it looked like the trees turned into grass, we were going so fast. When we slowed down, it was evening, almost as though we traveled an entire day. I looked down at this log house and saw this woman as she nervously waved to her husband and two children, who were heading away from the house in a buckboard, being pulled by two horses. I looked at this lady who was wearing this old-fashioned and dusty dress with a dirty white apron tied around her waist. She looked around, she looked up, and finally took a deep breath before she headed into her house. The next thing I remember, I'm standing on the ground near where she's hanging up her laundry. The interesting thing is that like these visitors, I, too, was wearing a hood. I remember she looked around one of the sheets she was hanging up and screamed. It scared the heck out of me. I remember being told to run after her, so I started and everyone else followed.

I finally caught her as we ran through her back door. And just as I did, the others were all around me; they took her, kicking and screaming from my arms and into a sectioned-off room. What was interesting was that this lady looked and dressed like the women in *Little House on the Prairie*, except that her dress was dustier and her hair looked like she pulled it into a bun and tied it as best she could. Another thing, she looked like the wife of one of the first settlers in the Old West. She was white. I remember looking around the room and finally looking in the direction of where this lady and the visitors were. One of them came out and thought (said?) that this lady's condition was worse than he thought and that she would have to leave with them if she was to live. I remember that this lady had no choice in the matter. She was going, like it or not. Next thing I knew, I'm looking down on her house and watching her husband and children as they come home. I can tell he's calling for her. As I watched him head into the house, the children looked up and saw the vehicle we're all sitting in, but by that time the husband was in a panic because he couldn't find his wife. We watched as they looked all over the area around the house and into the forest a bit. But then we left. I don't know what happened to this lady after I was dropped off. I don't even remember if she was in the same vehicle I was in or not.

I have had a couple of other dreams, or experiences, depending on how you want to look at it. One was that I was standing in the mountains, on top of this really big hill. At the top of this hill was a bunch of rocks, really big rocks. I remember looking around and seeing what looked like drawings of people on one of these vertical rocks. As I walked closer to them, I noticed that the drawings were actually pushed a little bit out of the surface of the rock, almost as if someone was buried underneath and the rock surface was placed over the body. The drawings looked as though they were standing up. I remember being told that these were the ones who had died while on earth and were therefore buried here.

Another dream was when I was older, about fifteen or sixteen years of age. I remember waking up in a big hallway. It was weird because it wasn't lit with light that I was used to, it had a sort of pinkish and orange colored glow. I asked what I was doing there and was told that I had to act as a translator between one of the visitors and a human official. So I was led down the hall to a big room where I was introduced to this human official, who was dressed in a military uniform, and the visitor. I find it interesting that they decided not to use their mind communication on this person. While they sat down, I was left standing. I remember it taking a bit longer than usual because it seemed that I hadn't spoken the language in a long time. I don't remember how the meeting went, but I do remember that the military official was surprised that another human was the translator.

I also remember, when I was about twenty-one years old, being taken to what looked like the Badlands in South Dakota. I remember feeling depressed because it appeared that I somehow ticked off these visitors. I was forced to sit down on top of this little hill overlooking this small flat area ahead of me. I recall that I was crying and begging them not to make me stay. I remember this voice in my head that sounded angry, telling me that when I got used to looking at these visitors, I would then meet them. I watched as a ship or shuttle came down and landed. Some small people came out, and when I wiped my eyes, I saw that they resembled the person on your book. I remember being so scared that I was shaking. I don't remember what I said or did to warrant this action. I don't recall if I had ever gotten used to seeing them.

When I was about nine years old my family moved back to Boulder. It was where I grew up. We left for a small vacation from Boulder and my dad, since my parents had just gotten divorced. We finally moved back in 1982. The house we moved into was alongside of a mountain. The backyard was a wooded area. I remember being awakened in the middle of the night, and finding that it was very light outside. I thought that I overslept and the sun was already up. I headed downstairs, I don't know why, but it seemed natural at the time. I noticed that the back door was open. Not thinking, I closed it and locked it up again. I turned around and then headed toward the bathroom where I knew someone was waiting for me. As I turned the corner, I noticed a little boy standing in the bathroom. He was my age, and he was about as tall as I was. He seemed to be as surprised as I was; neither of us knew why we were there. Instead of going to wake up my mom, we decided to play tag. I remember that we made a lot of noise, but I knew that my mom wouldn't wake up. Finally, after a long time passed, and we both were out of energy, the back door opened again. I remember watching this little boy walk out the door smiling. Even to this day, I remember what he looked like at that age. I say that because I remember that he had been visiting me almost every other year. We grew up together.

I recall a sad day with him though. We both were standing on top of this hill; we were facing each other. He had said that we could no longer see each other. When I asked why, he said that it was because it was getting close to the time when we were to meet. I must have looked confused, because he said that we were to meet soon when I was fully conscious, and when I was in my environment. This conversation took place almost two and a half years ago, but to this day I haven't met anyone that looks like him.

The only person who didn't think I was insane was my youngest brother. I think that he believed that if he did not join my siblings in trying to undermine my sanity, he would see himself as insane also. The reason I say this is because one night, I woke up to the sounds of somebody in our kitchen. It sounded like someone was digging through our refrigerator. I remember that when I woke up, I was facing the wall. I was terrified. My mom wasn't home, she had gone with one of my sisters to the reservation to look for a house since my family was moving back. I decided that since I was the oldest, it was up to me to find out who it was. The only problem being, the phone was in the kitchen so I couldn't call the police to report an intruder. I turned around in bed slowly so I faced the door. I took a deep breath and opened my eyes right away. All I saw was a shadow that then took off into my brother's room. Panic had now left me because I thought my brother might be in danger. I took off after it and found that the window in the room was open. I looked up and down the street to see if I could see someone running away. After a thorough check of the house, I went back to bed. I decided, like you, not to mention the experience to anyone except my mom when she got home. Instead, my brother came running down the stairs all excited in the morning. He asked if I had seen the person who had jumped through his window the night before. He didn't recall that I was chasing whoever it was, but he wanted to know if I had seen someone run by the room I was sleeping in and run out of his window. When I asked him to describe it, his description was similar to mine, it looked like a shadow. When I ask him about it now, he refuses to believe that he had seen anything. Well, that's about all I remember in terms of encounter experiences that I've had to date. I'm almost sure that I have had more, but the memories have not yet returned. One other thing—my family now lives on the reservation. I was sleeping in the living room of the house that they have, when I woke up late at night. I opened my eyes and noticed a lot of blue light coming from the hallway, around where the staircase is and the side door. I don't remember anything else, just the blue light. I'm still not sure what it was or what happened afterward.

My mom raised my siblings and me in the traditions of our people. We were raised to believe that seeing spirits (ghosts to non-Indians) was a part of life. So we have never been scared when we see something that others have a hard time explaining. Many people of late have seen lights in the sky and on the road, but they have been so ingrained with outside society that they fear anything the authorities can't explain. My mom has told me stories of when she was young. She and my aunt had seen a UFO, my grandpa had tried to shoot it with his rifle, but the sound it made was almost metallic, almost like a reverse metallic pop. I hope you understand this. Another time when she was almost three years old, when a meteorite crash landed on the reservation, all of the people thought that a piece of the sun had broken off and landed here for a purpose. Everyone who heard about it came and a camp was made around the meteorite. A lot of people were cured because of the heat. I don't recall off the top of my head what diseases were cured, but everyone thought maybe the Great Spirit took pity on his people and gave them something to help with their pain.

We went back to the site where the meteor had eventually stopped. There wasn't a lot of it left. Some of the people had taken pieces of it with them. My family picked up as much of it as we could. We took the biggest piece to the University of Colorado's geology department I don't know what they did with it. I have no idea what we did with the pieces that we had.

Recently, the wave of UFOs has been rising again. I'm back on the reservation for a visit. I have seen only one since my arrival. Many people don't know how to react to them. Some are spreading misinformation based on what they see on television, that we will be taken for experiments, etc., etc. A few have reported that their cows have been found dead with body parts missing. Some more have stated that they see weird little people near the casino, which is located out in the middle of nowhere.

One more thing. Last night I was talking to my mom about your book and about some of the experiences I've had. I heard three loud knocks on the wall of the house, not on the door, mind you, but on the house. This was the third time I heard these loud knocks. However, there were only three knocks. I asked my mom if she heard them. She surprised me by saying it was probably my brother who was taking a shower at the time. Maybe he dropped the shampoo or something. I was shocked at her reaction, but what followed was three of the weirdest whistles or bird calls, or whatever, that I've ever heard. I had asked her if she heard what I heard. She said she hadn't and wanted to know what it was I heard, so I described it to her. I then told my mom that I wanted to go outside and see who or what was out there. Her reaction? "You are getting into that book too much!" I know that on my way home, I was trying to see if they were still watching me. I decided to see if I could use my mind to speak with them—it worked. I told them that I wanted to try and get over my fear and had asked if they would come back. They said yes, which would explain all the noises outside of the house, but it really doesn't explain my mom's reaction.

A LIFETIME OF JOURNEYS

I have forty, or so, years of experience with the visitors—my first clear memories are from about the age of five. I value your work very much. For the past five years I have been getting persistent direction to contact you, but I was a bit worn out with the visitor experience and ignored the advice. So, last week I was maneuvered into going to a bookstore that I never visit—a Borders—and there you were. After all these years of prompts, I expected to feel an immediate connection with you. What I felt is your credibility, your daunting competence, and that you are a complete (but comfortable) stranger to me. So much for anticipating their methods. What the meeting did do for me was to rekindle my will to endure and participate in the experience. I read *Breakthrough* as fast as possible and enjoyed the "trueness" of it and I am still being pushed to contact you. As I have nothing to contribute to what you are saying, I can only think that I am being asked to tell you my story and the conclusions to which I have come.

During my teens and twenties I was convinced that I was very afraid of birds, particularly of pigeons. Their gray skin and huge, flat black eyes terrified me and they made soft whispering sounds. Years later, when I was in my thirties, I saw the cover of *Communion* and realized that something was wrong. Pigeons don't have huge, flat eyes, and these pigeons talked to me. It only took a few hours to remember the rest. We met in the garage at my grandfather's house, and I would sit out there for hours talking with a tall, thin gray lady with a pointed chin who told me to call her Grandmother. Grandmother put me through a kind of schooling. She taught me to use my mind and senses for telepathic and other communication. She showed me pictures of the earth from space and little video pictures of things that would happen to the earth when I was about fifty. She taught me other skills that are just coming back to me now. I have strong telepathic abilities, perceive things without regard to the time in which they occur, and am beginning to recall instruction on moving through walls, etc. I also have a memory that we can "fly" without aircraft. Anyway, she was a loving, patient teacher. Her skin felt like very soft leather, and she had a pleasant but cheeselike smell. She quit coming when my mother objected to what I was learning, and did not reappear until I remembered her.

Last year I got a little more comfortable with the experience and asked the Grandmother to teach me symbols to use during the infamous night visits. The most productive symbol I asked for was "I want to remember." She showed me how to project a picture to them in which I place a bit of bright light in the center of my forehead and then draw two lines of light from that point, circling my head and meeting in the back. Some nights later I awakened to see a visitor standing against the closet wall near my bed. I had no sensation of fear of what was going to happen. He was just standing there for me to see. As I watched, he took a slow step backward through the closed closet door and disappeared from my view. He was about four feet tall, intelligent looking, and made specific eye contact with me as he stepped back. The message I heard was, "Now you remember." I loved it. After reading *Breakthrough* I am seriously thinking about resuming the communication.

JOURNEY INTO PEACE

I am very nervous in what I am about to do. I am a channeler for spirits, extraterrestrials, and Brotherhoods of Light and Love and Truth.

I hold channeling classes every Wednesday night, in my home, on spiritual growth and change. On October 4, 1995, at the group's gathering, I had three beings who came and talked about the light and the dark. The first being called herself Katrina. The second called himself Gork. The third being is Kabir, whom I channel on a regular basis, To my knowledge I have not channeled Katrina or Gork before.

All of this week, starting October 2, I have been slowly reading your book, *Breakthrough*. I have not read any of your other books, due to my channeling these beings of light.

Ever since I picked up *Breakthrough* to read, all I have heard is, "We are visitors and we would like to channel through you on Wednesday night to the group that meets at your house." I meditated on it and felt good about it. I gave my consent to let them come into me and speak. Katrina and Gork said to me that they were visitors and that they wanted me to send you a copy of this channeling session. I must say it was a very good session.

However, I did not sleep well on Wednesday night because I was worried about what you would think about the tape. I asked the Spirit, "What will I do if he starts asking questions seeking to find out if Gork and Katrina are really visitors who have been in contact with him?" I heard a voice answer immediately, "Have I ever let you down?" "No," I said sheepishly. I hear the voice say, "Do not worry, trust, I will know what to say to his questions." I said, "OK, I will send the tape to him." I still tossed and turned all night anyway and got up out of bed at 3:30 A.M. to meditate.

Whitley, your book is fantastic and it could not have arrived at a better time for me. All the incidents and contacts really hit home for me. I have been experiencing the dark side, fighting, kicking, cussing, and screaming all the way.

However, I do not have the vivid sight that you have. I have asked for it, but was told I would be afraid to handle it. All my life, I am forty-five years old now, I have had so many incidents to let me know that they are there.

When I was eleven years old I had a bicycle accident that almost crippled me for life. I was unconscious for a short period of time and awakened in the arms of a man who had picked me up off the ground and carried me to the top of the hill to my house. I startled him when I awakened. I jumped out of his arms and said, "Please do not tell anyone that this happened." He looked at me as though he was in shock and said, "But you were dead!" I was bleeding from my head, both knees, and elbows, and was very scared but calm also. I sneaked into my house and went into the bathroom and filled the tub up with water to soak my wounds. My mom came rushing in and wanted me to go to the hospital, but I refused to go and told her not to worry, that I was all right. She tended to my wounds and I just lay around in a daze for the rest of the day. The next day I had a baseball game, but I could not play, even though I had suited up for it. I hurt all over and was still in a daze from the wreck. A week later I was running and jumping and back to normal.

I did not remember anything that happened during my state of unconsciousness. I noticed that somehow my life had changed. I was a kid and did not think anymore about it.

I remember after the accident that there were certain incidents in my life in which visitors were letting me know that they were there. One day I was playing catch with a friend in my front yard and I felt a presence around me and I heard it say "I am from out there." I stopped and looked toward the sky and then continued to play catch.

In the eighth grade I saw my first dead person in a casket. That night I felt the presence of someone in my bedroom. It came to the side of my bed and then I watched the bed sink down like someone sitting on it. I said, "Get away from me," and threw the blankets over my head, peeking out every now and then to see if it was still there. It was not and I went to sleep.

After that night I would lie in bed, and as the ceiling light was off and it was fairly dark in my room, I would stare at the ceiling and think about the shadow world and I would hear a voice that would say, "There are other worlds, the opposite of this one." I would think of these other worlds as I would drift off to sleep.

In 1985, I started to awaken to a higher spiritual awareness and started to read books like *Walk-Ins*. I remember reading it and then started thinking about my bicycle accident and wondered if maybe I was a walk-in.

It was not until 1991 that I went through a regression that would give me some information about my accident. In the regression, I saw myself going upward. It was dark, then it was light. I could see all these souls standing around and I was waving to them as if I knew them. They were saying, "What are *you* doing here?" I passed them and went higher and found myself on a spacecraft.

I remember being scared and could not understand why I was there. There were people that were human looking and beings that were not. Then a being so beautiful came through a door and when I saw this being, I knew everything was OK.

He was about six feet tall, with long white flowing hair, white beard, and beautiful eyes filled with light. The love was so immense. I knew him, but did not know how I knew him. We talked with our minds and I asked him who he was. He said, "I am God." He asked me, "What is your name?" I said, "Bill." He said, "What is the difference, as Bill is God, God is Bill, it's all the same."

The next thing I knew I was in the man's arms at the top of the hill.

In 1995, I went again into a regression back to the bicycle accident. With guidance from a hypnotist friend of mine, I got pretty shook up over it and wondered if my mind was not playing tricks on me.

Most of the information was pretty exact as before, but there was added information that made me feel very uncomfortable. I passed the souls as before. I left the light and darkness was all around, and as before, a spaceship was there. There were these guardians standing at the door with ax-like weapons with long handles. Their faces were like a skeleton's and the bones were white and their eyes were black and I was taken inside the ship. Once inside the ship everything was as I said before. When I returned I saw the man carrying me up the hill to my house.

One question that I had always thought about was, why would any grown man pick up a small child off the ground after a bicycle accident, dead or alive, and carry them to the top of a hill? No ambulance, no police, no nothing.

Then in 1993, my worst nightmare came with a speck of hope. I was living on the beach in Florida. I had just moved back there from Orlando, and had come to live on the beach for healing myself. I had just gone through six months of hell with a Crone's disease attack. I had no medical insurance, no money and I had diarrhea for about two months. Thinking I was just going through a cleansing stage, I realized that it was my Crone's disease and that I needed medication quickly. I called clinics and hospitals and no one would see me right away and it would be two to three months before they could.

I had lost about forty pounds and was beginning to wonder if maybe I had AIDS. Finally, I checked into the VA hospital and they okayed treatment for me. So off to the beach, medication in hand, to heal.

I was doing private channelings and psychic readings to pay for rent and food. Then one night everything went into total chaos. I was in my apartment and was told to go lie down on my bed, that it was time for an out-of-body experience. They told me that my experience was not going to be like others I had heard about.

As I looked at my ceiling, lying on my bed, I heard, "Now relax, take deep breaths in and out." As I released, I was told that they would be sending in energy and that the Kundalini would be sending out energy. I could see within me and felt the energy coming in and going out.

I was told at this point to start holding my breath until they said to release and breathe. I became so scared, so scared as I lay there, thinking they were trying to kill me. I do not want to die.

I heard, "Bill, Bill, trust me, trust this voice, I will not hurt you." I calmed down and finally decided I was going to hold my breath even if it was going to kill me. At that point, I could feel myself detaching from all my lower chakras up to my throat. Then as if there was a portal to enter my mind, I popped from my body and into my mind.

They were all there, I could not see them but I could hear and feel them. They said, "You are now out of your body and into the consciousness." My body was just slightly breathing as if it were on automatic pilot.

I was told, "You are now in your conscious mind, turn, look upward, see that portal. Go through it." I literally popped through this portal and was told, "This is your subconscious mind." It was vast and dark. They said, "Turn, go to the next portal." I turned, popped through this portal, and there were all kinds of images of houses, a kid on a tricycle, people in bars, having a party, asking me what I was doing there. I was told this is the lower astral plane of your consciousness. They said, "Turn to the next portal and go through." I popped through it and the energy shifted, it was so holy, so quiet. They said, "This is the healing plane of consciousness, here is where everyone goes to heal their broken bones, cuts, anything that can be healed." Anything. "Turn and go to the next portal." I popped through and the energy was great here; they said this was the plane where miracles and magic take place and then manifest themselves on earth and become your reality. Next portal I popped through, I was in ecstasy. I lay in bed just going, "Wow, this is awesome." There was no end to this vast consciousness of light. Just no end and if there is such a place in heaven, just incredible. The next portal I popped through, there was Christ. This was the Christ consciousness. Next portal, the energy got stronger, this is the Buddha consciousness. Next portal, popped through, this is the consciousness of the Mother of creation. It was awesome. Next portal, this is the Father of creation.

All of this was great; however, I started popping through all these portals coming back down into my body. When I returned I could swear that every demon in hell was there, talking, fighting, yelling; voices all at once, it was dark and I felt so abandoned. First the beauty and the wonder of it all and now the depths of darkness.

For the next two weeks it was as if someone had a string tied to me and I was their private yo-yo toy. Pull the string up to the beauty of light. The energy was awesome and back down at the flip of the wrist to the bowels of hell. Sometimes, out of control and no one could help me. No one.

Then one day during this two-week period it was as if they were taking a break and a man came to my house wanting me to channel. I explained to him what was going on with me and told him I did not think I could channel. I was tired, confused and upset, scared out of my wits. He sounded desperate, so I told him I could try. We went to the backyard, sat down and the sun felt so good on my body. I tried to channel, but could not focus. Then suddenly I looked up to the sky and a beam of white light shot out of the sky into the third eye and I just started channeling.

After this two weeks of yo-yo playing, I was angry. I packed my bags and moved back to my home state. For the next two years, I could not decipher anything. Whose voice was whose; I fought them and I yelled at them. I tried everything that I had been taught to get rid of this evil that I was sensing, feeling, hearing.

I questioned my channeling; everything that was coming in was watched by me and scrutinized. Yet I still felt so evil, so alone.

Then I called a friend of mine, Janie. She told me about your book and said that you talked a lot about what I was going through. Your book helped me to understand that what I was hearing and experiencing was OK. Live in the darkness, embrace it, love it, integrate it into your being. Quit fighting it, go with the flow, we will not hurt you.

I feel so much better now that I read your book. You are not alone and many of us are experiencing the same thing in our own ways. There are so many of us who know they are here and that they are here to help, not hurt, even though at times it hurts so much. I have changed so much in these years. It truly is a miracle.

No one has ever been hurt in any way by any of the beings that come through me. Some egos have been bruised, but so many people have changed in so many ways. Thank you for being here. People like me need people like you. From the above incident, I did receive complete peace of mind.

CHAPTER SIX

GROUP ENCOUNTERS

And we'll all go together
to pull wild mountain thyme
all around the blooming heather . . .

"The Wild Mountain Thyme"
Old Scottish Ballad

Many Witnesses, Many Secrets

The letters in this group are characterized by the multiple-witness effect, even though the events described are as divergent as a woman buying cigarettes in a deli from a counterman from the beyond, to a vision of Helen of Troy that emerged out of a complicated multiple-witness encounter.

In one case, the multiple-witness part of the experience does not begin until the original witness is hypnotized as a lark by some of his friends. It's not an "official" hypnosis: these are simply friends playing around. What happens is totally unexpected, and illustrates one of the most incredible secrets of close encounters: the sensation of being *inhabited* by an intelligent form. I have had this terrifying experience myself and, unlike the witness who wrote the letter about it in this section, the feeling did not disappear. Over time, I simply got used to it. In the end, it seemed to wear out. I never had the sense that anything had actually left me. What I had to do was get used to the sense of this presence. Unlike my unfortunate correspondent, I did not feel any malevolence from it, and when I could no longer detect the presence of my secret sharer, I rather missed him (or her, or it).

Throughout this group of letters, there is also the suggestion of another vast world that in some sense seems to contain our own—almost as if earth is part of some larger reality that completely encloses it and all who inhabit it.

And then there is the city—the elusive golden city, city of crystal, city of light. As I reported in *Transformation*, I once had a vision of flying over such a city, immense, shining in the light of another sun, its streets hauntingly empty, its windows blank, its stunning size and beauty filled with a meaning that I could not articulate, except to echo other witnesses to this ascended Jerusalem: it is a place where the truth is known.

We have also included here one "military letter," of which we have received around seventy, although few as fully detailed or as movingly written as this one. Some of these letters must have described operations of a secret nature, and were generally written because the individuals involved, often along with their families, had become close-encounter witnesses as an outcome of their work. Often, these letters were quite frantic, and we communicated with some of these people personally.

However, we were never sure exactly what we were dealing with when military letters were involved. In one or two cases, attempts to authenticate specific operations failed, either because the operations had been removed from the record, or because the events described to us simply did not happen.

In general, though, military letters, like the one reprinted here, ring with authenticity. From them, as well as from my contacts with individuals involved in investigations for the House and Senate, and military and intelligence community contacts, I have come to the conclusion that most of the United States government and military know next to nothing about UFOs, even though there are large numbers of operational personnel, especially in the air force and navy, who have certainly encountered them.

However, there is also a very extensive group of activities that appear to be a response to an apparent alien presence, and an attempt to emulate alien technology. The funding for these activities is in part accomplished through the "black budget," without the knowledge of Congress and the executive branch, and partly through the defense industry. The origin of the private funding is obscure.

In any case, the visitors appear often in group situations, as these letters will attest, and the military is no exception to that rule.

They Went to Their Knees

I do not believe that an abductee experience is in my recent history. What is in that history, however, are at least two incidences of directly observed nonterrestrial maneuvering, airborne vehicles, the most recent of which occurred while camping and exploring old mine shafts in a mountainous area. After crawling into my Gila monster-proof sleeping bag at 2215 hours (10:15 P.M.), I observed a dime-sized (from my view) circular object in a static mode, with a white, brightly luminous intensity. During a period of an hour and seventeen minutes, this object emitted, in a random pattern, a small flare. Each flare was differently colored and accelerated rapidly, and disappeared, after traveling approximately twenty degree of arc. During this seventy-seven-minute observation period, the object expelled a minimum of 250 of these "flares" before rapidly diminishing in size and disappearing from my ability to observe it, either directly or with an eight by thirty-power optical aid.

The first incident took place in July of 1957. I was stationed on board an aircraft carrier as a Radar Systems Operator. We were on station, and accompanied by four destroyers. I believe each of these destroyers had three separate radar systems, while we had five separate systems.

In the early afternoon of this most memorable day (to me), I was operating the SPS–8 height-finding radar repeater located in the Combat Information Center, just forward of the superstructure, between the hangar bay and the flight deck. The SPS–8 repeater displayed a very characteristic triangular radar pattern, allowing the operator to determine the altitude of nearby aircraft. Very suddenly, the entire top third of the repeater screen "whited out." Initially, I thought I was the target of a large-scale practical joke, with the perpetrators draping a sheet or blanket over the SPS–8 antenna. Within a few seconds, radio communications from all four destroyers became loud and excited, then the off-duty officers and men in CIC ran out and onto the flight deck. My screen remained partially obscured, and the radio transmissions became descriptive of a very large cylindrical object located immediately above our carrier. Prior to successfully begging to be relieved of duty in order to observe this object directly, I ordered the redeployment of another radar system antenna. Eventually, all radar returned reflection of a metallic object.

Systems on all five ships reflected the existence of the object. I joined over 3,000 officers and men from the five ships to watch the sight of a dark gray, cigar-shaped object over 4,000 feet long and about 350 feet in diameter, mirroring our every manuever. Later, through triangulation methods, we were able to determine the object's altitude at 63,000 feet, and thus were able to adjudicate its dimensions. Dozens of officers and men, including both Protestant and Catholic chaplains, had assumed a genuflective position, with hands clasped in front of their faces, or raised on high, a most stirring sight indeed.

Our ship's photographer was able to take over twenty pictures through a 1000-millimeter telephoto lens, and this film was airlifted to CINC PACFLT before the film could be developed on board.

The object remained in position for about twenty minutes, then began a demonstration of horizon-to-horizon random course and speed maneuvers that were of an incredible nature. This display lasted about ninety seconds, to finally end when the object "snapped" out of sight and radar contact, directly over our ship.

The carrier captain refused a request from the ship's air operations officer to launch jets to investigate. He indicated that he was not going to threaten the lives of his pilots "a corps perdu." Most probably, this was the wisest decision, although several pilots had already volunteered for that assignment.

To complete the above story, I would pay dearly to discover the whereabouts of those pictures, and to be allowed a glimpse of their content.

"THEY ARE NOT VISITORS. THEY ARE SIMPLY 'OTHERS.'"

Ever since I was little (I'm twenty-five), I've always had dreams where I flew up into the sky with the help of "fairy godmothers" who came down from the sky. There were always three of them, but I never saw their faces or knew their sex. I also remember being taken to a maze that was made of either glass or crystal, to find my godmothers, who were at its center. Finding them was the only way I could return home.

Also, when I was little, I recall someone telling me I came from the stars, that I was a star in the sky until it was time for me to be born, then I'd fall from the sky. I know for certain that it wasn't my mother who told me this; she always told me I was found in a cabbage patch. I guess I was the original Cabbage Patch Kid, ha-ha!

I've had only one experience with an owl. Up until I was eight, my family lived in California. The children in the neighborhood would play in the court I lived in, after dinner. I remember one evening my mother was with us. We were playing hide and seek when someone shouted. A large gray owl flew toward us. It was flying very low. It then turned around behind us and flew back over us, then disappeared. I don't remember what happened after that. I recently asked my mother if she remembered the owl. She remembered it as vividly as I did. I thought an owl in the city was very unusual.

When I was eleven, my mother, stepfather, and I moved to the Sierra Nevada foothills, where we lived for six years. My flying dreams continued. During those years, I saw a few UFOs and experienced what I believe was an ET encounter. I'd now like to describe the vessels I've seen:

The first one was a ball of yellowish-white light. There were two lights revolving around it. One light was a pure white light revolving in a vertical direction. The other was a red light revolving in a horizontal direction. The UFO hovered above the trees, then zigzagged very rapidly across the sky a few times, then disappeared. This happened in the middle of the night. I went back to my room and fell right to sleep, totally unafraid.

The second UFO sighting occurred after friends from across our canyon reported seeing a group of flying orange triangles in the night sky. The night after they called us I was outside searching. I didn't see anything out of the ordinary, but I went out each night. On the fourth night, I saw three orange triangles flying in a triangular pattern over our house. Once again, I wasn't frightened.

The third UFO was the most interesting and beautiful thing I've seen in the night sky. Again, my mother also saw it. Mother and I were outside on our front porch. Being summer, it was hot and we couldn't sleep. We saw what started out as a bright light that started to get bigger. We realized it was coming closer. It stopped and did nothing but hover for a few minutes. Then the most beautiful "spikes" shot out from it. Some were short and others were long. At first the spikes were the same white as the "inner core."

They then changed colors, starting with red then going through the colors until they turned purple. The UFO rotated slowly for a short time. The spikes then turned white again and then went back into the ball of light. It started to move toward our house. I was thrilled!

We went out on our back porch and waited. What we saw was fantastic! At first we heard a hum and saw the craft fly over our heads. When I say that we heard the hum, I should also say that we felt it as well. We didn't see the bright light. Instead, we saw a large black disk. It looked like the "hole" you described. As it flew over us, we saw what appeared to be lights, or perhaps windows, all along the outside of the craft. After it reached the edge of our canyon, it went up very high and changed to the bright light again, then darted away. My mother and I went inside, told my stepfather, who never believed in UFOs, and then I calmly went to sleep.

I'll now tell you about what I perceive to be my visitation. There isn't much to tell, as I don't recall too much at this time. My bedroom had been painted, so all the furniture was removed. The night before it was to be put back in I decided to sleep in my room, instead of the living room. The windows had been open for days so that paint fumes weren't the cause of my experience. Sometime in the middle of the night I heard a noise that sounded like a deep-throated growling. I woke up at the sound, and saw the most astounding light shining through my window; it was a pure light. I've been trying for years to describe it, but gave up. Anyone who has seen this type of light doesn't need a description. At first I was so scared that I couldn't move or scream. When I finally got out of my sleeping bag I ran to my mom's room. She heard the noise and got up, thinking I was choking. She saw no light. I went back to sleep calmly, but in the living room.

The next morning my mom and I searched our property, but found nothing. We tried to recreate the light using everything from candles to cigarette lighters. We joked that it was Bigfoot smoking outside of my window. In keeping with our warped sense of humor, we nailed a small handmade sign on a post outside my window, hoping the aliens would do something. (What, I don't know.) Our sign? "$E= mc^2$."

I've also had a recurrent dream of somehow being told that the world was about to end and I was one of many who were chosen to go to another planet to survive. It was *my* job to gather others. Not a lot of people were willing to listen. Those who did were rewarded with survival. We were led to a cliff at dusk, and a big spaceship of some kind came down to rescue us. The next thing I remember is walking through the streets of this place with the other survivors. The buildings were small and square, one-story stucco or clay. We were led to a building where two robot-type beings told us that the planet was where we'd be living and we would be the rulers, guided by them. Then we were brought back to Earth, because it wasn't time to stay there. We had to gather more people for when the time came to go back. Then I woke up. I've had this dream three times. I've always had dreams that, when I wake up, I'd swear on my life were real. I've had dreams that I've traveled to places, and I knew deep down that I actually did go places. I have several theories about this.

Several other members of my family have seen UFOs. My

mother and father saw one at fairly close range. They were driving in the country to a cousin's house. When they reached the end of a wooded area, a large lighted disk rose up and followed their car until they reached the end of the woods. This was witnessed by my mom's cousin, who was looking out of his bedroom window at the time. When my parents left and got to the edge of the woods, the UFO again followed their car from one end to the other. My mom says she'll always remember the hum. My parents reported their experience to the local police, and their story was in three newspapers.

When I first started reading *Communion*, I remember saying out loud, "They are not visitors. They are simply 'others.'" What that means, I really have no idea.

MORE OWLS

I'm writing you because I don't know of anyone else who can help me with my problem. I've never found anyone involved with the UFO community who did not have an obvious agenda, excluding yourself, and you don't strike me as the type who would tell me how the "grays" are connected with the Illuminati or Oliver Stone or some other such nonsense. I hope that you or someone you know might be able to help me understand what happened to me.

I just want to tell you beforehand that I don't know how much of what happened I accept or not. I have no answer for it, no explanation, not even a private one. It is completely out of my paradigm, so to speak. You might have a better chance figuring it out than I do.

I had left my house late at night because I was supposed to meet a girlfriend of mine on the main street of the town I was living in at the time. I had called her and she said that she would be there.

I was sitting on a bench in the middle of Main Street—this was about three A.M.—and I saw these giant owls flying back and forth across the street. They were the kind which I think are called barn owls, and they had huge wingspans, along the lines of four feet or better. Perhaps they get that big, I don't know. Anyway, I watched these owls fly around for a while. They would fly from the edge of one building, swoop across the street, and perch on the edge of the building across the street. They did this for a while, and I just sat there watching them. I can't remember my exact thoughts at the time, but I remember being rather uneasy. This may have been due to the fact that some of the local teenagers had threatened me recently, though. The town is a redneck place, and I'm definitely not a redneck.

The girl I was waiting for never showed up, and around four o'clock I went home. When I spoke to her the next day, she claimed to have no memory of our meeting plans. (She recently called me, completely out of the blue, and I asked her about it again. She still doesn't remember me calling her at all.)

I forgot all about it for several months. It just seemed to be one of those weird little things that happen from time to time, I suppose.

Then one night, in an apartment where I was living, in the next town, one of my friends mentioned that he had done some hypnotism in his college classes. This of course led to the "Hey, I can't be hypnotized, man!" line of conversation. I was one of the people who claimed that I couldn't be hypnotized. So my friend said he would regress me. He asked me if there was anything I wanted to find out about—he jokingly suggested that I see if I was molested or something—and for some reason I asked him to regress me to that night. He agreed and my other friends sat back to watch the floor show.

He hypnotized me, and I can tell you that I was out. It wasn't like being asleep, though—it was more like when your mind goes blank sometimes and you'll answer questions or whatever and not really know what you're saying. Do you know what I mean?

Anyway, he regressed me, and that's when it got strange. I'll describe it as I remember it from that night, or at least how hypnotism said I remember it, which I'm not quite sure I entirely believe.

I was sitting on the bench, and I looked up, and one of the "grays" poked his head around the corner of a building across the street. It was a bizarre movement—not like he leaned around the building, really, but more like he was on roller skates or something and just moved out from the building. I looked at him and he looked at me and I was completely terrified. I remember the feeling—not like nightmare horror, but the sort of terror you'd feel if you looked out the window one morning and saw your car floating in the air. A feeling of "wrongness," I guess would be the best description.

I then looked up at the sky and there was a hole in it. I want to make this clear—it wasn't a silhouette, it was a hole. I could see the edges, and stars curving into it. It looked like the popular image of a black hole, though I know enough about astrophysics to know that wasn't anything like a true black hole.

And then I was somewhere else, but it wasn't regression. It was happening real-time. It wasn't a memory.

I was standing in the middle of a red plain. The ground beneath my feet was dust—like what I might imagine moon-dust would be like. There were no rocks, no chunks of anything. Just this fine powder, but my feet did not sink into it. It's hard to explain that.

I appeared to be in the middle of a street. There were large, tan buildings running up and down this street, in all sorts of strange configurations. They were not elaborate at all—they were just very angular. On the whole, they looked like Spanish missions, if those missions had been designed by Salvador Dali. They were made of some crenelated, metallic material that, on first glance, looked like adobe . . . but it was obviously some metal. I asked myself what this was, and I got a reply. It wasn't from outside my head or anything, though—it seemed more like I just *knew*. This voice whispered, "Malachite" to me. Now, I know what malachite is, and it doesn't look anything like that. I have no idea whether that's important or not.

The sky above my head was white. Not bright white, or cloudy white—it was more like the sky glowed, like it had some innate property of light. Do you know what I mean? It was almost as if I could see the edges of the sky. That is also rather hard to explain.

On the street were dozens of the "gray" creatures. They appeared to be gliding back and forth up and down the street. It actually looked like the sort of thing you'd see anywhere, people just walking back and forth. They could see me, that much I know, but they gave off this feeling like I was sort of distasteful to them. I felt big and dirty and ugly, and I realized what a homeless guy might feel like in the middle of Beverly Hills. It wasn't a pleasant feeling. They glided around me, studiously not looking at me as they went along. There was no sound, but I could sort of feel them talking in my head to one another—like a radio picking up television audio, I think would be the best analogy. I got the haunting feeling that I was hearing their souls.

And then I felt a tap on my shoulder. I turned around, and

one of them was standing right there. He was another "gray," but I have to make myself clear. They weren't gray—they were white, like paper. They seemed to be made of plastic, like one of those bendable action figures that you can buy at the store. They didn't seem to have any sort of "structure," as far as I could see, in terms of skeleton or muscles or anything like that. They didn't even seem like living creatures, more like animatronic puppets.

The one that had tapped me on my shoulder was looking up at me with what seemed to be an awkward attempt at a smile. It horrified me somewhat because it looked like his mouth had just curled up in his face, not like muscles moving at all. He looked like a cartoon.

I didn't know what to say or do at this point. I was looking at a little white alien or whatever, standing in the middle of a street somewhere that I didn't know. But I remembered that, whenever this happened in movies, the guy always acted really awed by the aliens. That may sound silly, but I decided that I wasn't going to act like I was in a Stanley Kubrick film.

"Where am I?" I asked him, though I didn't hear any sound. He said something along the lines of, "You are in the Ninth Configuration." A note: I checked this out and it's the title of a book by William Peter Blatty. The book isn't very good, and the Ninth Configuration in the novel is the sequence of molecules that created life on this planet. I'd not read the book before, and I don't know what it means, if anything.

"How did I get here?" I asked him. His reply still frightens me. "You're always here," he said. "There are worlds within worlds; there are movements within movements."

I don't know how to explain what that does to me, now, a thousand or so miles away.

I was irritated by this response. "What the f**k are you talking about?" I asked angrily. I don't know why I was angry. Maybe it was because I was frightened. "Who are you?"

He said, "My name is—" and then it gets weird. The name started with an N, but then it ceased to be aural or explainable with an alphabet. It was geometric images in my head, I guess you'd say, but not Euclidean or fractal geometry. I don't know what it was, but it wasn't anything I'd ever seen before, and I can't even tell you what it looked like.

We spoke for a while, and I don't remember what was said at this time. It was a lot of very Zen-Buddhisty stuff, I remember that. I had little patience with Eastern philosophy at that point, and it frustrated me deeply that he spoke like some hokey sensei from a kung fu movie.

"Why can't you just come out and say it! I hate this Grasshopper bullshit!" I said, referring to the *Kung Fu* TV series.

"Our minds don't work the same way," he said. "I'm not doing it on purpose. I can't . . ." and then he seemed to make some decision. He moved toward me, and everything went black.

I woke up and my friends were standing over me, obviously very worried. I had no recollection of what had happened after I went black, though I did remember what happened before it.

What they said happened is beyond my idea of rationality, Mr. Strieber. Please bear with me on this. It sounds as crazy to me as it must sound to you. I have little tolerance for New Age sort of stuff, usually, and I really think Shirley MacLaine is a nut. But here it is.

They said that I jerked once, hard, and then went tense, like all my muscles had clenched up in the chair. They said that my posture changed, and when I spoke, my voice was different—lower, more raspy. It was not me speaking, Mr. Strieber. I swear on anything holy it was not.

What this—being? creature? said to them I don't know. They said that it mentioned something about "Hawking" and that he was close. I assume Stephen Hawking, the physicist. But they were rather scattered about the whole thing, understandably.

When I woke up, I was horrified. Not because I had had dealings with aliens, but because I was convinced that I had multiple-personality disorder. I think it's safe to say that trance-channeling is not in my nature, and nothing like that had ever happened before. I was in a complete and utter state of shock, but they convinced me that we should do it again, so I agreed to a couple of nights later.

The exact circumstances of what took place next are hazy in my mind. They recorded it, however, but I've never listened to the tapes. I do remember that I returned to this place, and that I was angry with this being for taking over my body or whatever it was that he did. He didn't understand that it was wrong to do it without my permission, and I made him promise not to do it again unless I agreed. He did, and I did, and it happened again that night. As I say, I am hazy about that night . . . but not the next night, when we did it for the final time.

This third night, my friend regressed me, and I was immediately "taken over" by this thing. At this time, my physical state was much more violent—every muscle in my body tensed up, and I was lying across the chair, flat as a board, my neck resting on the top and my buttocks on the bottom. This time the being wanted to show us something. It asked for a pen, and my friend gave it one, and it opened my eyes.

I'm not really sure how to explain what it looked like when it did that. It was as if there were too few dimensions in sight, like you woke up one morning without depth perception. I was conscious of something else in my mind, like a blank spot that I didn't understand. One friend swore that when I opened my eyes, they were bright blue, as opposed to the dark blue that they usually are. I would put this down to imagination, however.

This being took my pen and piece of paper and tried to draw something, using my hand. It was very clumsy with it—it felt frustration at having too many fingers that moved the wrong way. It held the pen between the index finger and the middle finger, clenched painfully tightly, and started drawing on the paper, and then *through* the paper. But it was just a bunch of crossed lines. It looked a lot like a honeycomb, actually, but I don't know what that means.

My friend was very concerned and tried to pull me out of the hypnotic state, but the being wouldn't let me go. I was awake here, but it was like I couldn't get the goddamned thing out of my head. I was screaming, he was screaming, and finally we got the thing to leave. But it left me with incredibly painful welts on my back, as well as some permanent back problems now—nothing serious, fortunately, but certainly a reminder that I didn't imagine it all. Since then I have not been hypnotically regressed at all.

Mr. Strieber, I'm scared. After this happened, I felt like I had been standing on bedrock, and that it had dropped out from under my feet, leaving me floating in an ocean whose bottom I could not see. I felt as if everything I thought I'd been clear about for seventeen years—the things that people learn—were lies, and that *this* was what was important. But it scared me, and I wasn't sure that I wasn't crazy. I in fact tried to commit suicide because I didn't want to be a lunatic. I felt like the people that Anne Rice described in *The Vampire Lestat*, "one of those tiresome mortals who has seen spirits." I was afraid that I would end up in the street somewhere, ranting about the end of the world.

Well, it hasn't happened. I'm eighteen now, and I'm leaving in a month to go to art school to study computer-generated imaging. But I'm still frightened.

I'm a very, very rational person, and to this day, I don't know what happened those nights . Was it an actual experience, or was it some subconscious imagery coming up to the surface of my mind? I don't know that hypnotic regression is necessarily valid anyway. This hasn't changed my life in the sense that I think about it and act on it every day; it has only changed me in that I realize how much there is that I don't know at all.

So my question is this: Do you know of any other experiences remotely like mine? I've read all of your books (except *Breakthrough*, which I bought today but have not done more than skimmed through), as well as lots of UFO books, but none of them sound anything like this. I don't know anymore whether I'm crazy or not. I don't *think* I am, and if I am, it's not an outward crazy. It's something that I can live with, I suppose. But I feel as if I'm in the eye of the storm, and I want to know what's coming. So any help you could give me would be sincerely appreciated. I want to understand this as a physical thing, as well as a spiritual one—which is what it seemed to be. And I want to know that I'm not crazy, hopefully.

P.S.: I did get as far in *Breakthrough* as your discussion of brilliant children. I learned to read at the age of two or three and at that time my IQ was off the scale. It's come down in the past decade and a half—I'm now rated at 165 on the standard scale—but I was always a very smart child. The first novel I ever read was Michener's *Space*, ironically enough, when I was five. Probably means nothing, but hey! You never know, right?

P.P.S.: I would really like some sort of response, even if it's just a note that you got this. It would mean a lot to know that somebody's out there.

NEIGHBORS

My first experience happened as a fifteen year old. One summer evening, my neighbors and I were outside. There were lots of children and adults. We looked up and saw a cigar-shaped object like you described in your book. It moved with great stillness. The windows were like those on a plane. Of course, no plane could have been that low, low enough to see windows, nor as quiet. The lights underneath it flashed red, blue, and yellow, sequentially. We all watched it for three to four minutes. It then shot off horizontally. I'll never forget the total silence on our street. No one said a word. Everyone broke up into groups and went home. The next day we all felt that it might have been some weather balloon or something.

The next time was in 1969. By that time I was twenty-six, with two children. My home was ten miles from an airport, across the bridge from an Air Force base. My husband was Air Force, and I was a sky diver at that point. There wasn't much I hadn't seen in terms of aircraft. One night a bright light came into my bedroom window. I started to turn over and go back to sleep, thinking it was one of the police helicopters that sometimes hovered about when crimes were reported. But the light didn't subside, and being a very light sleeper I couldn't sleep. I got up to shut the blinds. As I did, I wondered why the light was still there, and why there was no noise. I was in half of a duplex house, so I thought perhaps my neighbor had turned on the backyard light. When I got up and looked out, the light was coming from above, not from the other part of the duplex. I ran to the back door in time to see the light pass over my side of the house. It was sort of sliding along the ground and fence.

To this day I don't know why I did what I then did. In my pajamas and barefoot, I grabbed my keys and got in the car and started to follow the damned light at 2:00 A.M. The craft, by that time, was visible and it had rotating lights on the bottom that were circular. There was no noise. I followed it down some streets and right into the parking lot of a Kroger store. It stopped at treetop level over the lot. The only other lights were from the parking lot. I got out of my car with my hands and armpits pouring sweat. It was in front of me about twenty to thirty feet at tree level height, without any sound except a slight whooshing sound. It yawed back and forth with a delicate balance. It was near the phone lines. The big floodlight-type of light went out, but the rotating ones stayed on. I know this sounds crazy, but I looked for a rock to throw at it. I wanted to hear something solid, can you understand? I just wanted to bring reality into it. If I'd had a rock or was a pitcher, I could have hit it; it was that close. I sat on the hood of my car for fifteen minutes and watched it. I was evidently too stupid to be afraid, but I was terribly excited. It then went up slowly, and then shot up out sight.

It didn't even occur to me until I got home that I had left my two children alone, with my front door open and no driver's license or purse with me. I didn't remember any of this in the morning.

In fact, I forgot it all until two weeks later! At that time, I had a visit from a lieutenant from the Air Force base. Evidently I had called him that night when I got home. If he hadn't come to my house to verify the report, I don't know if or when I might have remembered it. Other than you and that lieutenant, I've never told this to anyone.

COUNTERMAN FROM THE BEYOND

In December 1984, about eleven P.M., I left my office after working late. The two building maintenance workers, Harry and Tim, were closing the building for the night, so we walked out together. Just as I reached the corner, three globes made of red light, about the size and shape of eggs, suddenly appeared in a circle around me. They floated around me rapidly, at knee height. Something in my head told me to stand perfectly still, that to attempt to walk through the circle of flying red lights would be very dangerous.

I am a large woman and sensitive to pranks, so I looked around to see if someone was playing a joke on me. I thought maybe Tim or Harry was using some sort of red flashlight to confuse and frighten me. I looked up and saw both of them standing across the street and called out to them, "What are you *doing*?" Both of them stared back with their mouths open in amazement.

Soon the lights vanished, and I continued on home with a sense of relief. I remembered that I didn't have any cigarettes and that I'd promised to bring home a quart of milk, so I headed for a nearby deli. All the counter guys at the deli know me; I go there all the time, but that night there was a new guy I'd never seen before. He was very tall—at least six feet three or four. His appearance was striking: he looked like he may have been a combination of African, Oriental, and Nordic blood. His skin was very pale with a yellowish cast, and his hair was negroid, with tight, kinky curls, except that it was the color of wheat. He had a broad mouth and very large, remarkable eyes, the color of gold coins, almond-shaped and slanted. He was dressed in the white uniform of a deli man. I was the only customer and we were alone in the store.

I put the milk on the counter and asked for my brand of cigarettes. I mentioned that he must be new there since I knew everybody else, and he smiled very gently and looked into my eyes with a searching gaze. When he gave me my change, I noticed that his hands were very smooth and his fingers very long. He seemed to be looking beyond me into the snowy night when he said, "It doesn't happen very often, but when it does, it can be very scary."

I looked at him and tried to figure out what he was saying. "What do you mean," I said, "the snow? That happens all the time—that's not scary." He looked at me as if to say, "Don't you know what I mean?" and then we both laughed. I had the crazy feeling that we both knew exactly what he meant, but I wasn't prepared to acknowledge it just then. I grabbed my paper bag with the milk and cigarettes and left, hailed a cab and went straight home. By the time I got home, I was ready to write off the whole weird business as imagination.

The next day when I went to work, Tim and Harry cornered me in the hall. "What were those red lights around you last night?" they demanded. They were very agitated; obviously they had been waiting for me to come in to work and the suspense was killing them. I was completely thunderstruck.

"You saw them too?" I said excitedly. "What did you see?"

"I saw you standing in the fire in the snow," said Harry. "The fire was going around you like this" (making a circular motion with his arms) "very fast!" I was determined to go back and ask the new counterman at the deli what he had seen. I wanted all the witnesses I could get.

When I went back to the deli, he wasn't there. I know the regular crew of countermen, and I asked them about the new guy who had been there the night before. "There ain't no new guy," said one of them.

"No, there is," I insisted, "you know, he's the tall one who looks sort of Afro-Asian? He was here when I came in last night."

"I was here last night," he replied, "and you never came in. I would have noticed because I didn't leave the counter all night long."

I began to shriek, "I was here! I bought milk and cigarettes from a tall man with golden eyes! I did! He was standing right where you are now; why are you denying it!" I was practically grabbing the poor man by the neck. Then I realized how ridiculous the whole situation was, and we both began to laugh.

BROTHERS TO THE UNKNOWN

If my memory is correct, the year my experience took place would have been either late 1961 or early 1962. It was during the winter months. I would have been close to nine years old. My brother Steve is twenty-one months younger than I am. Steve and I shared a bedroom. Very early one morning during this time period, I woke up. It was still pitch black out. All I can clearly remember is that it was way too early to get up for school. I now can't remember if Steve had awakened first, or if I had. I was somewhat surprised that he was awake also. Whether I had to go to the bathroom or get a drink of water or what, I still don't know why I woke up. I just don't remember. I do remember asking Steve if he was awake, since I saw movement from his bed. As I lay in bed, I heard a clap from Steve. Then he made another clap. After he repeated this a few more times, I irritably asked him why he was clapping. He said he was trying to catch the dots in the air. I was looking toward him in his bed, and I explained to him that there was nothing in the air, and to stop clapping. He insisted that there were colored dots in the air. Directly up, I saw thousands of these colored dots floating in all directions in our bedroom. These "dots" were bright, as if they were glowing. They were roughly the size of sequins, except that they were totally flat. These dots were in several different colors: red, blue, green, and yellow. I primarily remember the red dots and the green ones. The dots would make formations, and the one formation I remember very clearly was the triangle of green dots that floated and then went through one of the bedroom walls. I asked Steve if he saw the green triangle go through the wall, and he claimed he did. At the time, I thought he was just agreeing with me.

This experience lasted about twenty minutes to an hour. We were too fascinated to think of the time involved. After a while they thinned out and slowly diminished. Steve and I were too excited to get back to sleep. We both wished they'd come back, but they didn't.

Uninvited Visitors

Five years ago I submitted a report to the Aerial Phenomena Research Association about a sighting that occurred on August 1, 1965. In the report it says August 2, but in talking to my fellow witnesses we now know that date was August 1. In submitting this report about this UFO, I have always had to leave out the beings that I saw; they were not humans, I saw them at a distance of 120 feet, and then one individual about eighteen inches away. From what I can tell, you seem to have a good description of them. People regard me as some kind of a nut when I describe what I saw, so I've had to keep quiet about "them."

When this craft lurched toward the three of us on our dock at the Palo Duro Club, my first thought was to take evasive action. The craft was rapidly approaching, and I was preparing to dive off of the dock into the lake, twelve feet below. The dive might have been fatal; the lake being low, I would have wound up diving into water one foot deep and would have been killed or gravely injured. I had my foot at the front edge of the dock, which was about two feet high. I felt that I could dive into the water under the craft and hide in the weeds in the lake. Running eastward off the dock seemed fruitless; the craft was moving rapidly that way and I knew I couldn't outrun it. With my left foot on the top of the two-foot high wooden front edge of the dock, I began to push off and up so I could dive over the side of the dock, but my plans were to be changed by "them."

Suddenly, as this brilliant light bored into the three of us, I involuntarily sat down on the bench seat that ran almost the length of the dock. I was in a very uncomfortable position with my left foot elevated, but I couldn't move into a more comfortable position. I had to sit that way for the entire duration of the visitation. There I sat, with my weight on my left side and hip, and my left foot elevated upward. It was darned uncomfortable, but I was unable to do a darned thing about it.

Joe Walker was to my left and Tim, I think, was standing behind the railing separating the seating area and the rest of the dock. Joe and I were seated and Tim, I think, was standing behind Fred to his left. The ship began to let down, and by this time the very bright light had gone out. Now, as the ship came down, and it couldn't have been more than 125 feet away at that time, I began to see the upper part of the craft. Before, I had seen only the bottom, equipped with the "searchlight." Incidentally, it made a kind of whirring sound as it came down, like machinery running. When I saw the top side of this craft, I had a tremendous urge to wave.

Outlined in only what I can call "portholes" were humanoid figures. I could see the upper body parts of these beings, head, neck and arms, hands. They looked small, and as to coloration, at this distance it was hard to tell. The craft emanated a kind of cobalt-blue light, a gorgeous color I thought at the time, and the beings were in silhouette, with the light coming from their right and my left. They did appear sand-colored to me, but it was hard to tell. I was very frustrated to find out that I couldn't wave, or move, or anything.

I tried to say something to Joe Walker about the proceedings, but didn't, as I could move only my eyes. In front of our eyes was the wonder of the world, and I couldn't say anything to Joe or Tim to share verbally this experience with them. I couldn't run or wave, and I became highly frustrated. This frustration would soon be replaced by fear.

I was seated on the rightmost seat on the dock, and the craft chose to move about thirty to forty feet to my right. At this time, it must have been only a couple of feet off the lake. There's a kind of boat landing to that side of the dock, and as I watched the craft, now to my right, I began to see "things" flowing off of the craft. It reminded me of the way soldiers would have disembarked from a helicopter, but instead of going straight down from the edge of the ship as normal soldiers would into an LZ, these beings were coming off at an angle to the ship above the ground, and were moving past my vision to the right. They were like shadows, and were very quiet. Now it got very scary. I heard and felt "something" behind me. I heard Joe cry out in panic, and I knew how he felt. Figures began moving to and from the ship, again to my right, and it seemed like a continual procession. I got the impression that much of the time, these beings who were moving back and forth were not in contact with the ground. If they had been, there would have been splashing sounds as the ship was over the water. They didn't fly as birds would, but seemed to levitate in their chosen direction, which was toward us on the dock.

I saw the "ruler"; it made a noise somewhat like a blow-torch being pulled through the air. It also looked whitish in color, and reminds me now of those cold light sticks that some kids carry on Halloween, but it was smaller. When I first saw it, it looked like it was moving independently through the air. Apparently, it was being used on Tom, to my left. The hand that was holding it was very dark, and I couldn't see the figure that was standing behind me to my left. Apparently, the person who had the "ruler" had finished with Tom and I was next. The ruler and ruler specialist moved behind me, and I dreaded what was to come. My head was level, facing straight ahead, but suddenly my head was pulled sharply upward, and I looked fully into the face of something or someone. I saw this thing holding the ruler, and though I couldn't say it, I thought out loud "Is this all right?"; I got the mental answer, "Yes it's all right." I was relieved, to say the least, because if it hadn't been all right, we would have been in a world of hurt. The person holding the ruler looked down at me, and I felt everything was okay. Looking up, I saw this individual with the big eyes, but I thought they were goggles or dark glasses like the Ray Ban glasses that the jet jocks wear.

This ruler was placed on my forehead also, but I can't tell you how many times. Did you get the impression that it was a very sophisticated thermometer? It was also so dark that I couldn't determine his coloration, as that was about the last thing on my mind. I saw only him and the bright "ruler" that he held.

When this individual was finished with me, my head was placed back to its original position. I have searched my mind, and I know I wasn't taken aboard the craft. I think, though, that Joe might have been, but I could make very little of the comings and goings of the creatures as they flowed on and off the ship. It was dark, and the lights were dim in the craft. I have peripheral vision, but really couldn't see much.

After awhile, they were finished with their research (that's got to be what it was), got all their crew back on board and headed for my aunt's party, which was winding down on top of the caprock. That's where my cousin Angela saw and attempted to tackle the craft, when it landed about seventy yards from the cabin. My mother also saw the craft, and walked part of the way toward it with Angela, but Angela charged the ship, leaving Mom behind. Angela told me that she had very good feelings about the craft, that it emanated love and friendship.

Angela was so excited about the sighting that she was all over us when we came back up the hill from the lake. No one wanted to admit to even seeing anything other than a bunch of lights, and even then the memory was very hazy. Her persistence seemed to somewhat trigger our recall. She kept at us, and we began to recall much of what had happened. As the years go by, I can recall more and more detail. Your description of the "ruler" triggered my memory of it. I can now definitely remember that when I wanted my head to stay, it was tilted rudely upward and to the left. I was forced to confront, face-to-face, the "horrors" that had been stalking behind me and was greatly relieved to see that this person was friendly, and really not as ugly as one would imagine an alien to be. Compared to "ET," he or she was a real doll! We even had a gallon jug of beer on the dock, and I wonder if they drank any, as ET did. I guess the most prominent features were the eyes, so large and black that, for that reason, I thought they had to be goggles or monster contact lenses shielding the real eyes. They were just not in proportion to the rest of the body.

You know, we have a joke in my family: My late Aunt Laura was known for throwing some really great parties. We've had some very unusual people show up at them. How many parties have you attended where some of the party crashers arrived by UFO?

"I DON'T WANT TO TALK ABOUT IT!"

In 1975 we built a house out in the country. We saw UFOs all the time. My second son dreamed that he was taken to a base for UFOs. My daughter and her friend talked to a UFO. Most notably, my oldest son, who also is fascinated by the lore of UFOs, and I, were standing at the end of our front walk watching the sky about midnight. I believe we were looking at the Neried meteor shower, dating this event to August of 1976. A UFO of very large proportions arose from a field about a mile in front of us. It rose slowly and quietly until it was directly over our house, and therefore us. Suddenly its lights went out and it disappeared.

On reading your book, I realized my irrational behavior. Fascinated as I've always been with UFOs, I remember turning and running into the house so that I could get to a second-story window to get a better view. I always considered this very stupid behavior until I read your book. Now I realize I may have done nothing of the kind.

The third incident that I've always considered as a "close encounter" occurred on the night of January 27, 1977. On this night, my youngest son's birthday, my oldest son (again) and I were in the library talking. All the other children were asleep except the aforementioned daughter. She was reading in her bed. It was approximately 11:00 P.M. At least I remember that was the time. I had told Mark that it was time for us to go to bed. He had already gone upstairs and was preparing for bed. I started out of the library, and my attention was drawn to the window in the foyer. I remember my intention of going over and looking out the window. (I always look up when I look out.) I started to take a step when there was a brilliant flash of light. It lasted a very long time. Before I could move and when it was over, my son exploded out of his room at the top of the stairs, and my daughter came pounding down the upstairs hallway toward the stairs. They were both demanding to know the same thing, "What was that?" I walked the approximately ten feet to the window and saw nothing. We went into the kitchen to make cocoa, and to discuss the happening. This is what we discovered: we had lost upward of an hour of time. Where did it go?

The only person in our family who had seen people was my youngest son. All his life (he is now eighteen) he has been visited by "three wise men." I asked him about it the other night, and he was as reticent to discuss it as he always has been. At first he wouldn't answer my question, "Did he still see them?" Finally he shrugged and said, "Yes." I asked what they looked like, he fidgeted and finally said, "They are the three wise men, okay? Just leave me alone, I don't want to talk about it!" And that was the end of that. I was astonished when I realized I had also seen three elves (?) one night in late August 1968. I was recovering from a stay in the hospital, and enjoying the quiet of my in-laws' house while they were in Florida. Three little men came and poked at me to wake up and turn over on my back. I kept saying, "Go away. I'm sleepy." They ran up and down my back. I said, "Okay, I give up. I'll turn over." They peeled back the air like the page of an extra large book, and took me to see a house. When it was over, I asked the purpose of the exercise. They said, "So that you will know it when you see it," and they were gone.

I have a child that is the product of a pregnancy that disappeared at five months. For the disappearing pregnancy I did not go to the doctor. I "knew" that it would disappear. How I knew, I don't know. And I may be pregnant now, the product of a fascinating "dream." I sincerely hope I am not, but if I am, I am.

On February 22, 1988, I had a very strange dream. I dreamed that I had to go see Uncle so-and-so to see if it was all right for the child to come. So I went to Uncle so-and-so (I don't remember a name). He lived very high up in an apartment building. He lived in a little one-room bed-sitter, very small, very messy. He said he had to talk to somebody to see if it was okay for me. He looked like a little gnome, with a bald head, a sparse fringe of gray hair, and large twinkling eyes. He smiled easily. He went through a door at the back of his room where there hadn't been a door before. I waited, then giving vent to my natural curiosity, I went through the door. I was in what seemed to be a very old mansion. I wandered around and saw a school room, a library with a rose window, and a dining room with a very large dining table. What this has to do with it, if anything, I don't know. I was standing looking at the rose window when Uncle found me and exclaimed that I had found my way in. He said that it was all right for me to come in. I was to follow him to another room. This room looked very much like a doctor's examination room. There was a male there, and two females. I don't remember a whole lot of what happened next. I remember them bending me over a table so that my bare bottom was exposed. I don't remember undressing. They introduced something into me, and I felt a tremendous pressure. Then they started to put, like, electrodes on my bottom and my spine. They told me they were going to do it. I said to them, "I just don't want you to hurt me." They said that it would not hurt, but they wanted me to tell them what I saw. I don't know; that's all I remember. When I woke up and started to walk around, I felt like a virgin on her honeymoon. It came back with a rush, clearly and concisely.

I have found one other person with a memory of a similar place as mine. She remembers only the place which we call "The Planet of Light," or "The Crystal Planet." I was telling her of my memory and she said, "Wait, don't tell me. Let me tell you what it looked like." It was the same.

I want to tell you that my "impossible" pregnancy produced a daughter who looks very much like a modern-day Ishtar. When she was first born, I asked the doctor if she was Mongoloid, because of the slant of her eyes. At twenty-two, she had large slanted green eyes, and her beauty has been compared to that of Helen of Troy.

CARAVAN TO . . . WHERE?

I am thirty-nine years old and grew up in New York City. The first incident that came to mind while reading your book occurred in the summer of 1983. My husband and I had rented a house outside of Woodstock, NY. One night we both woke up startled by an extremely loud, whirling noise.

I said, half jokingly, "I think the spaceships have landed." I remember that the room was very bright, even though it was the middle of the night. My husband went downstairs to investigate, and when he returned he said he thought it had been a whippoorwill. We turned on the porch lights and looked down at the front yard, where we saw a large bird swoop through the trees. There remained, however, a feeling of unreality about the whole incident.

Another unsettling event happened in the summer of 1970. I was visiting a commune near New Paltz, NY. There were about twelve of us sitting on a grassy hill having a meeting. Suddenly, everyone got up and headed for his car in the driveway. We found ourselves sitting in our cars, revving the engines and realizing that we didn't know why we were doing this, but we were unable to stop. Then we all drove off, one car behind the other. I was with my then boyfriend in his van. One of the cars in front fell into a ditch alongside the road. The procession stopped; the guys got out, helped push the car back on the road, and we all drove off again. I remember being in a trance and watching the scenery as if it was a movie screen. The next thing I recall is the sky becoming very bright. It was already a sunny day, so the brightness had to be very intense. There was an electrical quality to it. Then I perceived what seemed like the sky opening. A force was coming toward me. The last thing I remember was finding myself outside of the car, lying alongside the road, trembling uncontrollably, in fear. My boyfriend was also lying alongside the road, but he wasn't afraid. I had always thought I'd had some sort of religious experience, but since reading your book I don't know what to think.

The more I read of your book, the more weird memories

come to mind. It was the summer of 1969. I was going home from work, and in the middle of the Seventh Avenue subway rush hour crowd I saw a little man about four feet tall. He had a huge head, but it was the quality of his skin that first caught my attention. It didn't look like human skin, but more like plastic or rubber. I knew he wasn't human. I tried to follow him with my eyes, but he quickly got lost in the crowd. No one else seemed to notice. This disturbed me; I thought I was seeing things.

It wasn't until I almost finished your book that a series of memories came to me which gave me goose bumps and chills. When I was about four years old I suffered from "night terrors." I would wake in the middle of the night screaming. The pediatrician explained to my mother that it was just a bad habit I'd outgrow. It was my dreams of that period that came back to me. I would have these recurrent dreams that I was looking out over our apartment window on West End Avenue, and see spaceships hovering over the Hudson River and the Palisades. The dreams were always very vivid and powerful. The ships were distinct with bright lights. I had completely forgotten those dreams, even though I've had them since childhood.

From the Top of the Hill

I've had a number of adventures with our "visitors from space," six of which I have so far pinpointed. The first one happened when I was five and a half years old, in May of 1933. There were apple blossoms and spring flowers and I was in the yard. We lived on a small farm in a somewhat remote area. I was watching my baby brother, and my mother had ordered me not to go over the hill or near it, and not to leave him alone. I watched him for awhile and then, for no good reason, sneaked off through the orchard and went over the hill, which was the top of a small ravine about 100 feet SE of the house. There was a rather steep bank down and a small stream/creek maybe twenty feet beyond. We children loved to go there, but my mother was afraid we'd fall down and get hurt.

This time, there was a UFO there with four "people" on it, at the bottom of the hill. They raised a shiny ladder for me to walk down, and then helped me aboard. I was terrified, because if my mother found out she'd beat me for leaving my brother and going. The "people" were very nice. I don't recall too much except that they put me on a table and lowered something that looked like the top half of a coffin, with wires and stuff, over me. Then they helped me up and let me go. When I got to the top of the hill my brother was almost at the end of the path through the orchard, looking for me. My mother came out of the door screaming. I got an awful beating. I couldn't tell her why I'd disobeyed her and gone. Always, after that, when we went down there to play, I expected to see a big round silver platform with a shiny ladder up the hill just sitting there.

The second incident was in July of 1936. The "people" were the same ones. There was a full moon, and I was on school vacation. They came in the middle of the night, and were very quiet so as not to wake my family. This time, though, my sister came. They were parked on the southwest part of the farm behind the outhouse and the chicken coop. My sister was afraid to walk where they wanted us to go because she was afraid of snakes. (There were lots of them, but none were poisonous.) This time we went up in the air and were gone most of the night, coming home just before dawn. I don't recall too much about this visit either, but down through the years I've wondered why my sister then went to bed, while I was so exhilarated that I stayed up to watch the sky and look at the moon, flying as high as I could on my swing!

The third experience was in August of 1941. There was a full moon again, and it was school vacation time. I was almost fourteen at the time, and I stood a head taller than their captain (the being in charge). I was then five feet six inches. The other beings were up to six inches shorter than he was. His name is almost on the tip of my tongue, but I can't raise it. They were very brown skinned, not Negroid, but just brown, and they wore something like brown coveralls, but not quite brown. I can still see the emblem he wore on the left side of his chest. His face was quite wrinkled, like an old prospector's, but more weathered than old. His hair was dark and he wore no headgear. The captain was a little solid and stocky, but the others were more slender. They were attractive and looked like humans. Remember that this was before World War Two. Now we'd call their uniforms "khaki." They were tied in at the ankles and they wore black boots. The uniforms had short sleeves.

One vivid memory of this through the years has been that they and my sister and I all stood at the back of the farmhouse talking, when my brother, who was a year younger than I, came out and insisted on joining us. If we didn't take him he said he'd wake Mom, and she was often a holy terror. The captain didn't really want to take him along. Why my little brother had come out too, I have no idea. We all went. The craft was parked in the same spot as the former time, though we'd had a lot of rain and it was pretty muddy. We all got muddy shoes, and had no way of explaining this the next day.

We all took off. There was that exam again, and the coffin-shaped thing was pretty scary as it came down like before. I was quite upset, because by then my claustrophobia was full-blown. But nothing hurt; there were no samples taken and there was no probing, etc. My sister sat and listened as the youngest-appearing one told her about how they know where to go. I remember asking a million questions of the captain, and I requested to go to the moon. My brother almost went insane! I don't think he'd ever been with them before. My sister could have been. He went so out of his mind that they had to sedate him and he missed all the fun. We did go to the moon. They showed us their base and their mother ship. The base was in a crater on the far side of the moon, well-lit and with lots of their people. It was enclosed in a force field for protection from meteorites and the cold. It was not in a hole in the ground, but rather was under an overhang of rock. They had a gadget that allowed them to pass through the force field, but we couldn't go in. We just sat in the UFO, looking out. The mother ship sat atop a leveled pinnacle of rock in the center of the crater; this was necessary for communications.

Then we returned home not long before dawn, flying slowly so that we could see what our area looked like. I insisted on going outside to see what it would be like flying alone out there. They agreed, and opened a small entryway. They gave me something to hold on to so I wouldn't slip or fall. I remember handholds and a rope or strap or something around my waist. Otherwise, there I was. It was wonderful, and at sixty-three I'm still thinking that maybe some day I'll take up gliding!

Then a strange thing happened. It was as if there was static electricity in the air, though the sky was clear and getting lighter; the moon was still full and about forty-five feet above the horizon. I was looking out and then down, and then things changed! Suddenly, there were no farms, no town, no roads, no anything, except swampy land with lots of reeds and very long grass, and flowers with huge heads nodding in the dawn breeze. We were about 200 feet up and we were swinging around to land and park. I saw lots of water with the foliage and a few very small trees. There were lots of large, funny-looking insects darting around, and some squawking came from below. I was really shocked, and I turned to tell them that we were in the wrong place! They immediately brought me back in, asking what I'd seen, and when I told them they were very surprised. I caught something that sounded like, "Too long ago for racial memory."

Then they landed and let us out. Things were normal, and we went inside to bed.

It was so hard getting up the next morning! We were to hoe the weeds out of the potato field behind the house. The bottom lower end was still muddy from all the rain. When the three of us older ones went out, we were deciding who'd work where, so we went to the lower end to see if that could be done at all. Imagine our surprise when we found two big circles in the mud! This was almost as surprising as finding all the mud on our shoes when we'd gone to put them on. My brother insisted that the neighbor boy had made the circles with his motorcycle, but you couldn't turn a thirties motorcycle in that size space; I knew that because I'd ridden with the boy a few times. Then my brother said, "an autogyro," but they had the same landing rig as they do today, and anyway, who'd land one in the mud in the dark? The circles hadn't been there the day before.

This was a mystery to us all, until a few years ago when I began adding two and two about this incident as I was remembering more. I discussed this with my sister, who recalled certain things, while I recalled others. We've no idea where my brother is now, and my sister and I have had a falling out, so there are no more conferences. Those circles in the mud were separate and very distinct, overlapping and possibly four feet across. The UFO looked like a round flying platform, almost flat but with a round observation room on top and windows all around that. If there were lots of lights as most reports say, I didn't notice them, though lights were used while landing. Nor were the "people" like any others I've heard of. I did see them one more time, though none of us ever had any memory of any of this until a few years ago.

I have now recalled a different visit. It was earlier in the summer of 1941. When I had a little free time, which wasn't often, I'd take a blanket and my current Zane Grey book and go down to the almost dry creek bed and sit under the pines and read. One afternoon while I was there, another UFO dropped down about ten feet away from me. The first one had been at least twenty feet in diameter, but this was about half that size. I jumped up. They hadn't come in the daytime since when I was four, but here they were. I wasn't at all afraid, but then out came these little gray beings with big eyes, forcing me to go inside. I didn't like how they acted, and I resisted with all my might, but they forced me inside of that thing. If you'd asked me later, I'd swear that I never had experienced an OBE, but I did then! I didn't want to go inside that thing, and then my soul left my body and watched them half carry my body inside, while I flew around looking for somebody to help me! Nobody was at home, and none of the neighbors paid any attention to me. I was forced to go back alone, wondering what was going on and why, who they were, etc. They were bringing my body back out when I returned. I waited until they laid it down until I got back into it. They were looking at me rather peculiarly before going back in and taking off. I never went to that place to read again, and I used to look at the spot in times that followed and wonder why I couldn't stand it anymore. Now I know.

I should mention that I never knew what those grays did to my body, since I was having my OBE then. I did have a sore spot on my right leg and a bad headache and sore abdomen. They seemed rather brutal; I'm glad I wasn't present with them.

Later that winter, the grays showed up again. I was engaged in lots of high school activities including Girl Scouts, and I often came home late on my bike. One night I came home too late, and my mother flipped. I was on my way home, at about 9:00 P.M., when they came again. Then it seemed that they just sucked my bike and me up into their vehicle, for they never landed. This time they laid me on a table. I protested as much as possible, which can't be much, when they coerce, as you know, and then they stuck something up my nose. It hurt like hell for a few days after that, too. I heard nothing from them until this was over, and then one of them either said or thought, "Now there's no evidence to find." There was something else too about "not being able to blame them," but I don't recall just how that went. They set me and my bike back down on the road and I remember thinking, "The least they could have done was to take me home so Mom wouldn't yell so much!" Boy, did she ever!

The last time I saw the beings I originally saw was before I went off to college in 1943. I was sixteen, and taking a speed-up course at the state university in July. They came at night again, to say good-bye. I know that I cried, but they said they were getting too old for that job and had other things to do, so they had to go home and were telling me good-bye. To my knowledge, I never saw any of them again.

They told me I'd do all right in my life, that I'd be lucky. They also told me there really was a God and that I'd find this out for myself, and that one day I'd be both rich and famous. I've been lucky most of the time, but not money wise. Parking spaces are there when I need them and I always hit the right aisles when shopping. I've never been injured but have had close calls. My judgment on what's good for me and mine is excellent. I have four handsome and smart kids who were born perfect, and my grandkids are the same. Some of them have 160 IQs. But I still scrimp and save, and I'm not famous. I did find God, along the way.

I can and have done healings that people have said were beneficial to them. I've been a "ghost buster," sending a number of them on their way, including my ex-husband, who died recently, very ill and confused. I'm sure he didn't know he was dead yet! I've seen many "ghosts," but never a "demon." I've preached, as well as heard, my share of sermons and given people many messages. Some are the result of my own psychic abilities, and some are from "spirits."

About the only being that I really fear is some insane, loony human out there. Even when the "grays" got me I wasn't scared, just furious!

There are two kinds of psychics, physical and mental ones. I'm not a physical medium; I'm a mental psychic.

Let me go back to that "primordial scene" on my last UFO flight, while we were returning home. The UFO people didn't cause me to see this I don't think; I just did. That scene came because I was excited, accepting, enjoying, and receptive to it. Today this is the most exciting part of that whole experience to me, that I psychically saw a scene from the ancient past of my home locale!

A Woodland Sighting

A few months ago I happened to be passing the TV during the day while a famous talk show host was on, and saw a book cover being held up that had a picture on it of something so familiar in detail that I had heart palpitations so strong that I had to sit down, and quickly. I had a few weeks of emotional struggle before I decided I wanted to read your book, to see if your story was in any way more detailed than your interview.

I am a social worker and counselor with teens, mother of three grown daughters and one teen son, and grandmother of five children. My parents were very religious, and I was raised in a very physically, emotionally, and sexually abusive home, all in the name of discipline and control. I married a minister when I was twenty-six, three years after my first husband was killed on his way home from work.

In 1973, we were visiting a family that lived in the woods and were building a cabin. They lived in a tent the summer they started. We were sitting outside the tent listening to our children's exchanges when we noticed a UFO across the wooded valley, against the mountains across from ours. We talked about it as we watched, or rather they talked and I listened with mounting fear for no apparent reason. It came closer to our campfire and no one seemed upset, including me, who actually was, then it disappeared. After dinner, at dusk and about two hours later, it reappeared about 400 yards away. I had the feeling that it knew we were there.

We saw a column of blue light, like crystal, faceted vertically, reach the ground. The children were off playing near the creek and our vehicles, a half mile to the right from where we were observing. Immediately I thought of the five kids and panicked, but as everyone else was so interested and not upset, I didn't dare become "unseemly." Soon it left and the children returned, natural and exuberant as always. We talked about the differences we'd seen. The ship from far away had seemed cigar-shaped, and to split as it sped away. Then, after dinner, it seemed like a fat disk and quite large, even though not really close to earth. It was then dark, where it had seemed that light glinted or twinkled before that had caught our attention from the other mountain, maybe ten miles away. It was pale blue-gray under it, where the column showed. It wasn't angular, but gave an impression of a softened, flattened triangle. We talked about whether it was half of the split "cigar," or a different ship.

When I saw the cover of your book on TV so suddenly, and I wasn't aware what the topic of the interview was, I felt a load of memory flood back to me behind my eyes. Not visual or total recall, but almost a tangible recognition of a person from one's past that you thought had died and was suddenly back in your life again.

I want to state a few incidents of similarity to your incidents:

There was a little white sylph who was a comforting one to me as I grew up. I also remember two or three times, from 1969–1972, as an adult, having nightmares and waking up with bruises on my upper arms and back and neck; I couldn't be coherent as to what the dreams were.

A Weather Balloon

In November of 1956, I was hanging out wash on the line at midnight. I had three young children then and was busy all day, so I would sometimes do the laundry at night, after they were all asleep. I saw lights moving near the horizon, and as they drew closer, I saw a craft with a dome-shaped light glowing from within it and brightly colored lights near the bottom, revolving in a clockwise direction.

It made no noise. It kept moving toward me at a steady pace, then descended until it was about two feet off the ground. I was getting ready to run out of the way when it stopped, moved sideways, and then rose straight up. It was almost as big as our house. When I looked up, I saw that the bottom of the craft had a circle of beautiful sockets around its outer edge. Each socket was made up of three gold rings, like giant donuts piled on top of each other, with a strange type of "Arabic"-style writing on them. There were twenty-four sockets, with a light on the end of each one. The lights were colored red, green, blue, orange, yellow, and white. These colors were repeated three times, in the same order.

The craft moved across the street, revolved in place a few times, then swooped sideways at an angle and was soon out of sight. At nine AM the next morning, I heard a report on the local news about the sighting of an unusual object in the sky the night before. The reporter said it was probably a weather balloon.

CHAPTER SEVEN

THE NINE KNOCKS

And so faintly you come tapping,
tapping at my chamber door,
That I scarce was sure I heard you—
here I opened wide the door;—
Darkness there, and nothing more.

"The Raven"
Edgar Allan Poe

Knocking Down the Walls of the Soul

In *Transformation*, I reported hearing nine knocks in three groups of three coming from the ceiling just beneath the rafters of my old and much lamented cabin in upstate New York. I have lost the cabin, but the memories of my wonderful days and nights there with the visitors remain.

Among the most haunting of these memories are the knocks, and they have also sparked a substantial response among readers. Again, in *Breakthrough*, I returned to the knocks. This time it was because a substantial portion of the population of an entire town had heard them after I wrote *Transformation* but before it was published. This, to me, was strong circumstantial evidence that the visitors were quite real—and led me to doing even more research about the possible meaning of the knocks.

Knocking is central to the initiatory process in Masonry,

and nine knocks in three groups of three are used in the initiation to the 33rd Degree, which is the highest publicly known degree in the Masonic tradition.

In the U.S., Masonry is in deep decline. Its obsessive secretiveness has both preserved it through many dangerous years and also doomed it to modern oblivion. Nevertheless, the Masons are still a powerful force, and certain elements of the Scottish Rite in Europe remain very much in tune with ancient ritual and lore, and understand the order's true age, its manifestation as the Knights Templar during the Middle Ages, and the tragedy that followed when the Templars were brutally suppressed by a French king greedy for their wealth.

The visitors told me once, when I was suffering from the severe persecution that led to the loss of the cabin, to go to the Masons and tell them that I came as the widow's son. I did this, but none came forward to help me.

I think that the visitors' knocking on our roofs and doors and windows represents a call to mankind to awaken and be transformed into an entire species devoted to the great work. As they announce themselves on behalf of resurrection and the value of the soul, they also announce themselves as Masons, and bring with them the suggestion that they are somehow connected to its ancient origins.

It is getting harder and harder to doubt that the visitors are as ancient as the Masonic Order has claimed itself to be, especially since it has been discovered that coded material in the Dead Sea Scrolls is virtually identical in structure to certain Masonic code, dating the Masonic tradition back to the time of the Temple at Jerusalem—far earlier than the fifteenth-century origin that scholars outside of the Order commonly admit.

Masonry should change its entire structure right now: it should become missionary and begin to proselytize its tenants, with the objective of returning this world to the fold of the master builder, so that every human being comes to understand the secret journey, and all men are Masons.

Masons from degrees beyond, it would appear, are calling to us in the dark. Listen to the knocks, one, two, three; and three in three, the nine that will be.

Three Knocks

Just finished reading *Breakthrough*, and was a little shook up by it. I had the "nine knocks" experience, and at the time I associated it with evil. Let me explain:

About fifteen years ago, I received a letter, one of those chain letters that starts "You must pass this on or die," or something to that effect. I laughed, tore it up and went about my business. That day I had two close calls. First a perfectly good tooth broke in half, and then came the knocks.

I was reading when the knocks came. There were three deliberate ones, a pause, and three more. By this time, I had had enough, and began to pray. The knocks were extremely loud, and gave me goose bumps.

I have had other paranormal experiences and don't spook easily, but this time I did.

NINE KNOCKS

I experienced the nine knocks twice in my life, both times when I was fifteen. The first time I did not react as if it was out of the ordinary, but the people with me did.

I was baby-sitting with a seriously ill child for the first time. It was the summer I was fifteen, and I told the people my friend was going to come with me. I didn't tell them it was a boy. After an uncomfortable hour alone, my boyfriend showed up, and he brought another boy with him. I was so glad I was not going to be alone with the chronically ill child.

At about 11:45 P.M., there were three knocks on the side of the house. It startled me and the boys. They talked as if it was a car backfire or sonic boom, and we continued with the gossip that so entertains teens. About twenty minutes later, three more loud knocks came, again on the side of the house. The boys jumped up and ran outside to see if they could catch whomever was doing the knocking. They each took turns knocking on the wall, trying to make the same sound, but they never could. Their reactions seemed so extreme that I believed at the time they were trying to scare me. I acted like it didn't scare me, so why should it scare them? Then they got angry, thought I had set them up, and left, mad at me.

Ten minutes after they left, there were three more knocks on the same wall. I was sure they were now trying to scare me, so I didn't pay any attention to it.

The next time was around November of that same year. I was baby-sitting at a different house. The knocks came again, on the front, higher part of the house. This scared me. I was alone and the kids I was sitting with were asleep. About twenty minutes later it happened again. This time I was near panic. I picked up the phone and called home. My dad answered and said he'd come to sit with me. Just minutes before he arrived, three more knocks happened. When he arrived he saw how frightened I was, and stayed until the parents returned home.

A Struggling Initiate

I think that you are very talented and brave to write about the aliens. I don't think I could even tell my neighbors about them, let alone the world. I am a student, and wouldn't even dare to talk about this in front of anyone.

I have become aware of them for the past few years. I now realize that they have been a part of my entire life. I am thirty and I have several vivid memories of them. I even have a scar on my knee that I think I received from them. Doctors will mention my "knee surgery" after seeing the scar, and don't believe me when I tell them I never had knee surgery. I also have scars where my tonsils used to be, but I never had them removed. I am scared of the visitors, but I am also very intrigued. I hate how they turn up in the middle of the night when we are all asleep. It wouldn't be so bad if they came when we're all up and about. I sense their fear of us, as well.

My memories are mostly like "real dreams." I wake up and am surprised that I am back at home. I feel kind of fuzzy, cloudy and electrical when I'm with them. At those times, I feel confused, and they act like I am supposed to know them. They never seemed evil to me at all. When I have these "dreams," I can touch, smell, and even taste the surroundings around me. I at first thought they were just crazy dreams, but I never have such bizarre and real experiences in my ordinary dreams.

I have seen the little gray beings and people that looked like normal humans, except their eyes are a little different somehow. I'm not sure why I think this. I normally experience them together, but I don't understand their connections with each other. Do you know if there's a relationship between them?

My whole family has had different experiences. Even my four-year-old daughter knows about them. I never talked about them in front of her, but she calls them "the brown people." I don't really know why she calls them that, since she says they look like that and have bug people (whom she calls "monsters") with them. She confuses the colors of gray and beige, and calls them brown. She told me they wore brown clothes, but I pointed to a grayish shirt as an example. She says they play with her on swings, and that I sleep on one of their "beds" while they watch over me. At first I didn't know if I believed her, or if she was making it up. Then one night my father turned on our backyard lights, and she hopped up in her bed excited and said, "My friends are here." That's when I started believing her. She also kept mentioning a name that I didn't recognize, but I later recognized it in a UFO book I was reading. That scared me! She now also has a similar knee scar, and I don't know how she got it.

It makes me mad to think that they are doing things that would hurt my little girl. She says they talk without moving their mouths, and she hears them in her head. She demonstrated by pointing between her eyes on her forehead. She says the "brown people" are nice, but the "monsters" are evil because they hurt her with a knife and scare her. Her description of the knife is like a rod-shape, with a light on its end.

I have experienced different things with them. First of all, I've never been in an airplane, but I have flown with these beings. I can vividly recall looking down out of a window-wall, and seeing cars driving by. I could see people through their windshields. They never seemed to notice us above them, but I never figured out why. The vehicle was traveling not so fast, maybe fifty-five miles per hour, and we were right above the streets and the power lines. Once in awhile, they seemed to rise up over a taller obstacle, and I could feel this movement. I remember thinking that I should have felt more than I did. Whatever I was in was huge!

There was this older man, or at least I had that impression about his age, with whitish-blond hair. He tried to explain to me why they were traveling slow and staying low, and how they did it. He explained that they needed to go to a certain area. It was much too technical for me. He also said they could defy gravity, and demonstrated this with his swinging arm. I never understood much, because it sounded so complicated. This man talked like he was glad to see me after all this time. I didn't remember him, but sensed that I must have known him at some other time. He told me to continue with my schooling for now, but he would let me know when it was time for me to do something-or-other. He gave me a little tour of the craft, and even showed me how to work the bathroom. He explained that there was no gravity in space, and how this worked for that purpose. The bathroom reminded me of a cross between a shower unit and a refrigerator. I know this must sound completely crazy! There were other people in this craft like me, but they all sat in chairs, zombielike. By this time, I was telling myself that I was either dreaming or losing my mind. He seemed to watch as if he knew what I was thinking and was amused. He wouldn't let me look directly at his face, and would turn away if I looked at him. He said he was going to "assign a tutor" for me, so I could start learning something. The last thing I remember is that he sat me down and pointed a silverish stick with a bulb-type end at my forehead. I saw bright lights, and the next thing I knew I was in bed waking up.

I thought this was a dream at that time, and told my family about my "weird dream." Then, not two weeks later, I had another experience. Every time I'd go to bed at night, I'd encounter this "woman" teaching me some symbolic-looking stuff. I never saw this person, because she'd stand behind me and wouldn't let me turn around. I would study this stuff for what seemed like hours. I'd wake up in the mornings feeling exhausted. This continued for about two weeks. By that time I was so exhausted I prayed that she'd leave me alone and let me rest. The lessons stopped. I would start to feel better, more rested, and then she'd come back for more teaching. I felt I was completely losing my mind. I remember that these lessons involved some sort of symbols, and that I was to say what they were. I can't remember exactly what I was learning, if that's what was going on. But after it was all over, I did experience a difference in myself. My school work seemed to have become so much easier. Even my grades showed it. I didn't have to study much, and I could remember everything.

A couple of experiences happened when I was awake, too. One time I had a very frightening OBE. I could see my body across the room, and I was scared. At that time, I thought, "If

I'm over here and that looks like me, then who am I?" I even thought I must have died. I was petrified. I felt I had to get back to my body, but I was scared of that too. It felt like I was floating. I went back after a few minutes, but I was so scared it would happen again if I went back to bed that I stayed up all night.

Another time, I had trouble falling asleep. I was lying there and then I heard a popping sound. I rolled over toward the door and saw a blue-green, swirling light hole where my door used to be. I was more amazed than scared at the time, and I remember thinking, "That's interesting." Another pop sound, and it disappeared. The next morning, my mother said she'd had a weird dream. She said she was led around the house by something or someone as she looked at everyone sleeping in their rooms. She said it felt like a lesson. She said that when she got to my room, she said, "I get it!", and then she was back in her bed. She said that, even though she looked at each one of us, one at a time, and ended in my room, it all happened simultaneously. I wondered afterward if my "light hole" was her, learning her lessons. She has experienced several more things similar to this one.

I never mention this to anyone outside of my house. I never even mention it to other relatives. I am an adult, after all, but am petrified of the dark. It makes me feel a little more normal to know that I'm not the only one in the world who is experiencing these things. I know several religious people who've mentioned this on their own, and describe the visitors as evil. I don't know why, but I feel angry at them. I might have had something happen to me to get the scar, but if they were evil, I'd probably be dead by now. I really don't think that a demon would treat a person with gentle respect, nor help them achieve a better understanding of themselves.

My grandmother, recently deceased, used to tell me her experiences. She said that, one time when my mother was a baby, she was doing the dishes and heard a series of nine knocks on the side of her house, near the ceiling. She called to see who was there, and nobody answered. She then went outside to see who was knocking, and a blue ball of light soared by her and around to the back the house. She said her neighbors saw it, too. It startled her, and the neighbors came around to see if she was okay. She said she never knew what it was, but had often wondered. At that time nobody ever heard much about alien encounters.

MORE KNOCKS FROM THE CEILING

I have had similar experiences to what you've had, and have always been a little "spooked," even though I know deep down that there's nothing to fear. I've had many OBEs, but have never gone all through them completely, with the exception of lucid dreams. I think I experienced them that way to ease my way into it without my being so scared.

I was amazed at the experience you had when you were in space looking down at earth. I experienced something very similar.

Most people would say it was just a dream, but I think it's another level of consciousness. If we could get over our fears, what a wonderful thing to experience at will.

The knocking, oh the knocking! Just a couple of weeks ago this loud, echoing sort of knocking, as perfect as yours, came down from the corner of the ceiling. I later tried to produce this same sort of noise, but couldn't. It was right above my head, the rapping was being directed from the ceiling to my ears. I was so creeped out that I never thought to count how many knocks. When I got up enough nerve to get up for bed, I went to the bathroom and, by God, I swear that the same knockings were reproduced on the bathroom ceiling, and again in my room when I crawled into bed.

I could write a book, so many paranormal things have happened to me. It is always when I am alone, or when people are sleeping.

I've had psychic experiences, knowing when someone's going to call, what people are going to say, when something's going to happen, etc. Some days, I experience nothing at all. On the days it happens, I'm in a sort of trance, feeling literally "high," like when you are walking and feel you weigh only a couple of pounds.

When I have my OBEs and I'm up by the ceiling, as soon as fear takes over I'm zapped right back into my body again. I think I am afraid of what I might see.

In the past couple of months I've had these two vivid

dreams. I might add that I don't believe they are dreams, because they are real. I woke up in the middle of the night and looked out the window, and lo and behold, there were hundreds of UFOs with lights blinking changing colors and zinging around the sky so fast. I had the strongest feeling that they were coming. Then there was the night that I actually remember waking up and sitting with my head turned toward the window. I watched a huge white light just sitting there, for maybe twenty seconds, then it sped off in the blink of an eye. After that, all I remember is waking up in the morning. I'll tell you, I wouldn't just go to sleep after seeing something like that!

Why must things like this be so sneaky and perplexing? I feel that since I've had these things happen to me, I am slowly changing for the better, and I am accepting difficult problems more easily. I'm more at peace with myself. It sounds like you are, too.

"When I Am Ready"

I recently moved in with someone who has some books on UFOs. They have always interested me. Shortly after starting one of the books, I felt a presence that seemed to come through the sliding glass doors and move across to the foot of my bed. Shortly after, it vanished. I only felt the presence, but my dog saw it moving, as she demonstrated by her snarling and barking, and visually following something moving across the room. I was pretty scared.

I have felt presences all through my life at various times. I've always thought them spirits, or something of that nature. It wasn't until I read your book that I thought, "My God, something has been happening all along that I have been clueless about." I have heard the nine knocks!

During the summer of last year, I was awakened when my dog started freaking out, growling and snarling and obviously frightened. I woke, and saw light outside my window. In and of itself, this would mean nothing, because I lived directly across from the Main Bar. Then, with my dog barking, I heard, *knock, knock, knock – knock, knock, knock – knock, knock, knock*. I got very frightened. At first I thought some drunken fool was banging on houses. After the third set of knocks, I went to the window and looked out, but saw nothing. The knocks came from high up on the outside of the house.

After going back to bed, it occurred to me that there was a meaning behind them. I remember thinking, "Three sets of three; what does that mean?" I forgot about the incident, until I read *Breakthrough*, and saw that many people in Glenrock heard the same knocks.

Last year, I also woke up to see a hooded figure that emanated pure evil. I was scared to the point of being paralyzed. After about twenty seconds it vanished, but I couldn't sleep for almost an hour afterward. Later, I talked to a friend of mine, and she had seen the same figure, and had the same feeling of evil.

When I was fifteen, I awoke because of a loud, masculine,

authoritative voice, calling my name. It came from right over my head and spoke my name with a certain urgency, but I didn't know where it came from or what it wanted of me. Eventually, I just went back to sleep.

The earliest and most profound experience was when I was about eleven. I had these recurring nightmares that didn't seem like dreams at all. They ended in a bizarre event.

I'd be asleep, and would wake up with a buzzing feeling and a kind of tightness in my chest. Every time this happened, I knew that I'd have the nightmare when I slept again. The nightmare, which I remember very little about, was of the feeling of a needle coming at me, and the feeling of being forced inside the head of this needle. This was the most terrifying feeling I've had in my life, even though it makes absolutely no sense. At the same time, during the dream, I kept repeating to myself, "This is impossible!"

The last time I had this dream, I woke up with someone else in control of my body. It was completely horrified by what it was experiencing, as it starting screaming insanely, while I was observing this all in a completely passive way. It ran out of my room into the hall, just as my mother and her boyfriend came out of her room to see what was happening. It saw Mom and her friend, and screamed again. It went down the hall into the kitchen, at which point it realized there was nowhere else to escape to. It started spinning wildly, while my mom came into the kitchen, grabbed and slapped me, and suddenly it was gone. So was the buzzing all over my body. I remember other things about the dream: colors, lights, and movement, but nothing solid.

I am wondering if I should have hypnosis to try to remember more about it, or if I should leave it alone. I have the feeling there will be more, when I am ready.

CHAPTER EIGHT

SEXUAL ENCOUNTERS

And Jacob was left alone; and there wrestled a man with him till the breaking of the day. And when he saw that he had prevailed not against him . . . he said, Let me go, for the day breaketh. And he said, I will not let thee go, except thou bless me.

Genesis 32: 24–25

Sexual Experiences

The encounter experience is profoundly involved with human sexuality and reproduction. There are all sorts of implications suggested, none of which make much sense in our ordinary understanding of the world.

What does make sense, though, is the impression that these letters leave of extreme intimacy, of *penetration* of our being that goes far beyond anything we can do to one another.

Among the things that make a certain sort of sense is the suggestion, repeated in one of these letters and commented upon extensively in the literature of the abduction experience, that children are somehow being created out of human sexual material and raised elsewhere, without reference to at least one of their natural parents.

If this is true, then mankind might already be extensively expressed into the cosmos, and the implications regarding family ties, raised by another of the letters, might, quite simply, be explained by the fact that a lot of people already have relatives who live away from the earth.

As bizarre as this seems, even more bizarre are the stories of a fetus brought to the end of the first trimester, then removed from the mother's body, only to be shown to her a few months or years later in the form of a child that is an apparent cross between a human being and some other sort of creature altogether.

Lest this be dismissed too readily, one of the apparent implants that has been studied was contributed by a witness who remembers a lifetime of sexual encounters and being shown at least thirty children, none of them completely human, who were the result of the use of his semen. The implant that he produced could not be explained by any known physical processes, and it was indeed extremely strange that this thing had ended up embedded in his leg, let alone remained there for thirty-four years. And yet the evidence is overwhelming that this is exactly what did happen. Does it mean that stories of hybrid babies are true? We do not know what it means, and perhaps we need to find out.

THE MAN WITH THE GOLDEN NEEDLE

In the summer of 1966 two events occurred which weren't dreams, yet they weren't reality as I knew it. However, when they happened I had no doubt that they were extremely real events.

My husband worked second shift and didn't get home until about 11:30 P.M. Normally, I would stay up until he came home, but with two children under five I was tired and went to bed about ten that night. I've never gone to sleep immediately and this night was no different. I laid in bed, facing the wall, somewhere in between the state of waking and sleeping.

Suddenly, as I stared at the wall it opened up and this big dog walked soundlessly into the room. It resembled a grayhound in form, but had reddish brown fur. It totally ignored me and walked into my kitchen as if looking for food and water. I could see through all the walls in my house and watched the dog pacing around the kitchen, but I heard no sounds of his claws clicking on the floor. I didn't know what he was doing there, but I was fearful that he would hurt my children. My two-year-old son had recently been bitten by a dog, so I was more fearful than usual. However, the dog remained in the kitchen pacing.

The wall remained open and soon people clothed in something resembling monks' white robes with cowls covering their features came silently through and surrounded my bed. I could see no faces, but could somehow sense both male and female. No words were spoken until one of the males said, "Have you got the golden needle?" I was quite taken aback as I hate needles. A female opened some sort of case and removed a gold needle with some sort of handle on it—almost like a screwdriver except it was a needle. Those were the only words spoken.

The male took the needle and began making a circle with it around the crown of my head—like a circle of dots. Then he told me, "Concentrate!" I didn't know what I was supposed to concentrate on, but I began concentrating on something so hard that I had a vicious headache. I could sense about twelve people surrounding my bed and knew that, somehow, I was helping them do something, but I had no idea what. It seemed to go on for a long time and the pain in my head got worse. Suddenly, the door opened and my husband came home. The people left through the hole in my wall as quickly as they came. Normally, if I'd been asleep, I never would have heard the door open. The next day the headache lingered.

About two weeks later I went to bed to take an afternoon nap and, again, I was in that state between sleeping and waking when a male walked through the wall. He wasn't wearing the robes, but I'm uncertain what he had on—if anything.

All I remember is staring at his eyes, which were a metallic gold and seemed to have pupils. I've no idea whether he was sitting, standing, or reclining on my bed. No words were ever spoken and no sounds of any kind were made!

The male, mentally, made love to me. There was nothing physical done at all! In fact, it seemed to me that I no longer had a physical body myself—that I was nothing but a mental creature and had no need of the physical realm, although I was vaguely aware of breathing. He took me somewhere and showed me a city that appeared to be underwater, but I couldn't tell for sure. I only know I wasn't supposed to breathe and really had no desire to do so. I felt more loved than I ever had in my whole life and wanted to stay there forever. My husband and children, my family, none of them could compete with the feelings I had in this place. I seemed to have stayed there for days, but I don't really remember all the things I saw.

The next thing I recall is a voice saying, "Breathe," but I didn't really care whether I did or not. I wanted to stay there, but was forced to return to my own reality. I was still euphoric, but sad. The male gave off the impression that he would come for me again, but I knew it wouldn't be soon enough. The whole event only lasted two hours of my time! For several days afterward I only required about two hours sleep a night.

SATANIC LOVER

I am barely able to write. Every year, for five years, I have sent you typed letters about my experiences. I had one good friend who lived in Texas and later moved to Roswell, NM. She and I never met, but the aliens put us together many times, along with other ladies in our age bracket (forties). The eldest lady was in her seventies. We all knew the aliens by name. Male aliens visited us all, either sexually or by abduction and rape.

One little who visited us was called "Abbe." He, along with a reptilian race of huge aliens, would abduct us. To some of us, it seemed like a spiritual experience, even though we all had the same "dream," "OBE," or whatever they did.

Four months ago, they did something to my right brain, and after that I changed and became bonded to them, almost enslaved. Then I found out that their god was not Jehovah; they worshiped Satan. At least that's what they told me after I asked about their god. Maybe not all of them worship Satan, but the ones I knew did. They attached something to my ankles, neck, body, and tormented me to obey them. The "nice aliens," I had thought so good, were threatening me into submission. After that, I had a psychotic breakdown. Since the mental health field doesn't acknowledge aliens, I was alone. I was termed schizophrenic due to the voices in my head. We all know that the aliens are telepathic, so convincing anyone was futile.

The voices, of course, quit after I was in the hospital, and I had no other symptoms of schizophrenia. I did have seizures and tremors due to the horrible things the aliens did to me.

Even though I had many witnesses to ships over the house, and a few friends even met the aliens in my house, no one else believed me. Proof is fleeting, and no one in the medical field wants to believe me. Now I am going to see a neurosurgeon to see if he can stop my seizures. (Later: he found nothing wrong.)

I wrote letters to the Pentagon not too long ago about aliens, and how nice some of them are. Now I can't even tell them that they are not so friendly after all. The planet that one of them was from was called "Calvus I and Calvus II."

They can be good or evil, and there are many types, but ask about their god and you will understand more.

While I was friendly with the grays and reptilians, my religion deepened and I became more religious. My marriage improved.

The little grays like to sleep with people. My grand daughter, who is three, said, "They talk to me, and then go into Mommy and Daddy's bedroom." I think that, if what the one gray told me is true, they are fooling us and are really demonic in nature. If so, then many will fall under their aura of goodness that they portray.

All my letters were usually positive. This is the first time I have ever sent anyone a negative letter, but since this happened to me, I know this is true.

Beware of soul-bonding when they enter your body. They will, in time, take you over.

It has been a long while since I wrote and there was a good reason for it. For awhile I felt I had the answers to the universe, God, and ghosts. The spaceships flew over my home in a regular bi-monthly display for me or my friends.

The bedroom visits were either interesting or scary, depending on which alien group visited. My married son saw one alien levitating on the road waiting for him, and my son shot him! I told him not to do this, but to my son, all aliens should die. My married daughter saw an alien take away her baby daughter, but returned her due to my daughter's pleas. My husband still has not seen any ships or had any remembered experiences, even though I have seen them over him while he sleeps. In 1992 the ships only flew low a couple times, but high and to the far north or south the rest of the year.

My bedroom visits became very sexual. This entity disguises himself as my husband and makes love to me. It seems vivid and real at the time, but the aliens did tell me when I first met them that they can control our thoughts, so who can tell? Many times on a weekend, the entity would lie down in my husband's bed, even his touch and voice sounded like my husband's! If I am dreaming, then I am doing this in an altered state. It is so real at the time. Enough said on that. My sexual experiences were always pleasurable, and I also remember other aliens making love to me. Now I feel like an alien whore due to these visits.

My husband and I are careful not to get me pregnant, but I have never used birth control. I have noticed discrepancies in time while we made love. One night while my husband made love to me, he removed himself from me and acted strange, trancelike, and started to fall asleep. He has no memory of this, yet I know we were making love and he stopped, never finished and I said, "Did you come?" He did not know! He fell asleep! (This has never happened before.) I remember a flash of light in the bedroom, so I believe the aliens interfered.

A UFO hovered over my home for several nights, and was seen by a waitress at a local restaurant. She is my only witness to that event. I also saw a jet zoom past it, and a laser (?) shot flew from the UFO to the jet. The jet kept going. On the TV show, *Sightings*, they showed a possible space battle over the earth. I have seen lots of secret things, but the aliens control how much I can remember, and sometimes I believe that they control who comes to the house!

Before I could send this out, the aliens abducted me again. I remember going to bed and all of a sudden I felt nauseated. It got worse, with pain in my stomach, then I felt a terrible pressure all over my body. I could not move; I felt like a lumpy sack of potatoes. Slowly I moved my right arm up to my head almost instinctively. I could feel my hair, but it was deeply indented. It was like they removed part of my head.

I felt a hand gently remove my hand/arm from the area. Again I tried to touch my head and again my hand/arm was gently put down. It grabbed me by the wrist. Other skin tests were done, maybe more. I remember thanking them for letting me remember. I don't remember any pain. It looked like I had a diaper on. Something was white and pulled up between my legs. They went up each foot with either their fingernail or some other sharp object and asked if I could feel it (I said yes). I remember my arms and legs floating upwards and thought that was great. At the time I felt like a child in a doctor's office. I was back in bed, the nausea subsided quickly. I was too frightened to look around right away.

GHOST FAMILY

When I was around three years old, my grandfather had an entire room of the house where in winter he spent his time building wooden lobster pots. He was in this room working, and the rocking chair began to rock by itself. Naturally, he was scared. He said a voice called him by name and said not to be afraid, and said, "We are here to look after the child." Grandpa grabbed me, as I'd just come into the room, and left the house and refused to go back until the rocking chair was out of the house. I used to constantly play paper dolls in this room. I remember that my favorite set of paper dolls disappeared, and we all searched for them and never found one piece. I had left these and many others scattered on the floor. All the rest were there; just those particular ones were missing.

When I was six, my parents and grandparents bought a house together. It was a huge mill owner's house on a hill. While they were working on the house, before we actually moved in, I saw a third staircase in the house. This staircase didn't exist. I never saw it again, and no one else ever saw it at all, so I thought I hadn't seen it then after all. At about the age of ten I was moved from the room I'd been sharing with my younger sister into a room of my own on the opposite side of the house, where my grandparents lived.

I'm not sure of the order of these things, but my sister and I both had the impression that we'd dreamed of a big, round spaceship hovering just above our backyard one early evening. We never thought we'd actually seen this, just dreamed it. Another night, I woke up and saw this enormous spider walking across my bedroom floor. Let me say that I have absolutely no fear of bugs or spiders, and have never deliberately killed one. When we find them in the house I'll even pick them up and move them to a safer place. However, this wasn't a regular spider; he was the size of a tarantula! Obviously, I couldn't actually have seen a spider that size walking across my bedroom floor! I had completely forgotten this over the years, and your book reminded me of it.

Around the same time, one early evening I was going to go upstairs to my bedroom. I was at the bottom of the stairs and looked up and saw what appeared to be a man looking down at me. I never saw his face. I ran screaming. My mother and grandmother checked and found nothing, but I was terrified after that to go up those stairs for years. My grandfather, coming down this same set of stairs to use the bathroom one night, insisted that he was pushed from behind down the stairs.

I was fifteen when my parents sold this house. I had such a horror of moving. I felt that I had to stay there, had to, and couldn't leave that house! The house burned to the ground shortly after we moved. They built another house there, but I still ride past there like I'm attracted to the spot!

Nothing else happened for several years, although I remember two other things. My grandfather once told me that he felt he was being "watched." I also remember a terrible preteen fight with my mother when I was around twelve, and telling her that she'd be sorry when "the lady" came and got me! I also had this definite impression, which is probably totally ridiculous, that although my mother was my mother, my father was not my real father at all. I even tried to get my mother to admit this!

I got married at age nineteen and had a daughter in 1969. Nothing at all was strange. One night I woke up. The baby's crib was next to my bed. I don't know what woke me, but it obviously had awakened her too. I could see her on her hands and knees in the crib, staring at me. I thought to myself, "What does she see?" I wanted to get up and go to her, and was unable to move. I found myself lying on my stomach; this is not strange, except that it's one position I never lie in! Although I was able to move my eyes, couldn't move because there was something holding me down. I felt that it was on my back, some heavy weight holding me. That's all I remember! My daughter was about eight months old then.

During the time following my divorce, alone with two small children, I felt myself attracted to a certain village, which no longer exists. It's only up the road from my house. There was a farmhouse there on a hill that my father said had been abandoned for forty years. I would drive right up to this house, but I felt this panic inside it. I would lock the doors and roll up the windows and say to my kids, "This is our house." My daughter was about three. One night this house mysteriously caught fire and burned down. My daughter sat right up in bed screaming, "Our house is burning!" She was hysterical. From then on, she has consistently predicted fires from her dreams, and is terrified about it. The most recent incident was just this week. She not only dreamed of fire, but woke up with a long, thin burn mark on each cheek. We thought maybe they were scratches, and I looked very close, but they were burns, both alike. The odd thing was that apparently my mother had the same dream on the same night, and saw it as a kitchen pan being on fire; she said Kelly was in the room.

One night I had a second incident exactly like the other one. It must have been in early 1974. Again I woke up in the middle of the night. I was alone in the room. I don't know what woke me, but there I was again, lying on my stomach! I knew what was going to happen, and then this weight was on my back again, holding me down. I argued and said, "No, don't do this!" and felt something quickly enter me, painlessly. Although I knew it had happened, I was left there, wondering, "What happened?" Of everything, this is the thing that scares and concerns me the most, because about nine months later I gave birth to a son.

This in itself wasn't a mystery at the time. I'd been dating someone for over a year and we had been intimate and there was no mystery involved. This was my third pregnancy, however, and I went four times to the clinic. I had all the chemical pregnancy tests, which all came back negative, yet I knew I was pregnant. Eventually, I went to have my fifth chemical test. The nurse was really irritated with me by then and said this was foolish, and I was not pregnant! I insisted. Again the results were negative. I insisted again that I see the doctor to be examined, which I did. He not only confirmed the pregnancy, but said I was several months along. He had no reason to give me for all the negative pregnancy tests.

This pregnancy was unreal. I was sick all the time,

ghastly sick, which I hadn't been with my two older children. I carried him three weeks over the due time, and had to have the labor induced. I spent much of that pregnancy lying on the bathroom floor crying and sick, and thinking I would die. I also had strange spells. Anything I looked at would seem to be outlined in glowing, flashing lights. This would be followed by numbness in my right fingertips, and a tingling that actually crept up my arm further and further. I could feel it inch by inch, up my shoulder, then up my neck and then I'd totally black out. When I snapped to I'd have such total amnesia that I couldn't even remember my own name for several hours! I was convinced I was having strokes, and my doctor simply said that he had never heard of such a thing, and that was the end of his interest. This continued throughout my pregnancy, and has never happened since, in thirteen years.

Despite my concern about these spells, I still thought there was nothing unusual going on. I had my second son in late 1974, a beautiful, normal son. A few months later something occurred that in no way I could pass off as any sort of dream! It was night, but not really late. I'd been up all evening and hadn't gone to bed, and was fully and totally awake. I'd left the baby asleep on my bed. I went in to check on him and found him asleep, glowing in blue light! My first thought was of some kind of light coming around the shades that might disturb him, so I picked him up and moved him. I moved him three times. I would put him in another spot to get the light away from him. No matter where, the light stayed with the baby. I felt no sense of fear or danger or anything except a kind of awed wonder. I then did something which later made no sense to me. I got up and left the room, leaving my innocent baby asleep and glowing in blue light. I had touched him, and the light felt warm, comforting, and unthreatening.

To think, though, that any caring and loving mother would have left that baby alone and left the room is so amazing to me.

There have been no other occurrences involving this son, but also I could never say that things have been "normal" with him. At the age of not yet two, we were amazed to find out that he could tell time and has this built-in directional signal in his head. My father was amused by all this. He'd put the baby in his car seat and let the baby direct him around. He seemed to always know where he was going. Another time, what really made all our mouths drop open was that my father needed to buy a part for his car (my son was older then), a part with a long number on it. As he was searching through his papers, my young son rattled off this long number. Dad found the paper and had my son repeat the number and he had it perfect. He had watched Dad write it down, but we could never get over his remembering this! We're convinced he has a photographic memory. Although he was raised totally right-handed, I discovered when he was ten that he was actually left-handed; just a lot of small but unusual things.

In the meantime I became pregnant for the fourth time, by the same man I'd been dating exclusively for several years. I had all sorts of pretty names chosen for this coming baby. One night while taking a bath, and this actually happened, something or someone said, "My name is Stacy."

Your book mentions telling lies you know are not true, and this occurred with me. My son was only five months old when I began to wonder why things hadn't returned to normal, and why I hadn't had a menstrual period following his birth. I dragged all three children with me to a clinic, and the first thing they did was a chemical pregnancy test, and the results were positive. I remember the nurse glancing at me and asking my son's age, and I told her five months. I then immediately set up an appointment to get an abortion.

My mother took me to the hospital for this. Now, none of this makes sense!

As these children were born twelve and a half months apart and I knew my son was five months old when I found out about this pregnancy, I had to have been newly pregnant, and the abortion should have been a simple one, yet I remember the doctor saying that I was four and a half months pregnant, and that the only abortion they could perform was the kind where the baby is actually torn apart. I was so disgusted at the very thought of this that I didn't return the following day for the abortion, and told my mother I couldn't go because my pregnancy was too far advanced.

There was also some confusion when this baby arrived. I went into labor three weeks early and went to the hospital around 12:30 A.M. The head nurse examined me and said I wasn't in labor, that the baby hadn't dropped yet, I wasn't dilated, and she even tried to send me home and I refused to leave. The baby arrived about a half hour later, in the examining room! Needless to say, I named her Stacy, because I felt that I absolutely couldn't name her anything else.

The next two events occurred about three days apart. Again I thought it was a dream. This time a woman walked into my bedroom. I felt that I was awake. I tried to get up because I was filled with pure terror. The woman wore blue, although I don't know what it was that she wore, and I felt like I knew this woman! I wanted to get up and scream or run, but couldn't move. I did shut my eyes, because I didn't want to look or see her. I felt her hand on my shoulder, and know that she said not to be afraid, and that she talked to me. The strangest thing then is that I asked her to let me see my daughter! Then a girl came in! The girl never came near me; she stood back and looked at me and said, "I would have loved you." I thought all this was some kind of weird dream and never told anyone. Three mornings later, my oldest daughter told me about a strange dream she'd had the night before. She said a lady and man had come into her room and she was scared and tried to get up, but couldn't move. She did mention she had shut her eyes so she couldn't see them! She said the man stood back and she didn't see him well, but the lady was dressed in blue. She also felt like she knew the lady, and the lady had put her hand on her shoulder and told her not to be afraid, and talked to her. This was too exactly like my own dream to be believed! How could we have had this same dream experience like this, three days apart, even having felt the same things? We had the same reactions, saw the same lady, and both felt she was someone we knew! I have to admit that at that point I decided we had ghosts in this house, and that this must have been the spirit of my grandmother. I pretty much convinced myself of this, although at the time, I didn't really think this was my grandmother. I just thought it "had to be."

This is the one and only time that I actually saw anything that didn't look quite right: What happened is that a woman came into my room and again I felt I knew her, but it wasn't the same woman that had come to me earlier! This one attempted to get me out of bed. This part totally doesn't fit in with the other things or with your book, because I refused to go! Not only that, but I feel I was actually off the bed several times, pulled off physically by this woman, and that I got back onto it under my own power. She said I had to come with them, and I refused to go. I was not paralyzed in any way. I also remember calling out Jesus' name during this ordeal, then getting back in bed. She would pull the covers off and grab my arms and shoulders trying to pull me up; I wouldn't go! During all this there was a creature in the doorway only a foot or two away. It never came near me, but just stood there watching. It was very short, and had a brown hood over its head. I do seem to remember glowing eyes.

I had left the door shut, as we always have a lot of cats, and I didn't want them in my bed at night. In the morning the bedroom door was open, the rug was scrunched up and my slippers were kicked around, so I knew I'd actually been off the bed. I still considered it a dream, but it scared me terribly! I remembered that a few years later when I saw *Star Wars* with the kids, and there was some little person in the movie wearing a brown hood, with glowing red eyes, then I remembered my "dream"; it looked so much like the one I'd seen in my room!

I feel no fear or terror, and furthermore, don't believe anything's ever been done to me in the way of experiments. I don't feel I've ever been taken anywhere. I have absolutely no such memory, and don't think anything like that ever happened.

I fly in my dreams, but it's such strange flying! It's such an effort to get going; it involves countering gravity with a total effort at pushing against it to get up at all. It's so hard to get up, then you go straight up, not with your arms out like a bird. Then there's no control; you can't stop going up.

JOURNEY OF A WISE WOMAN

The events/experiences I wish to tell you about happened in 1984 and 1985. In December 1984 in my mother's home one night, I was extremely upset and confused by my relationship with my boyfriend, and crying so hard that I couldn't think. I was alone. I finally heard a gentle male voice trying to calm me, and I began to listen and to grow calmer. I associated the voice with God, but I was later told this thought was wrong. I learned to associate it with "caring ones," whom I tend to call angels. The voice spoke as if there were more with him, and he promised they would be back. He asked me to "eat one meal a day and not to bathe." I honored this request for a week or so.

They returned to talk with me every evening at that time. I had no idea who/what I was talking to. I couldn't see them, but I could feel them and could hear them. I trusted them. There was so much wisdom and, at times, unwanted truth that it was scary coming from a voice that was not of my ideals, principles, or standards; they were completely alien. For those hours my mind opened, and I understood and comprehended a beauty that, to me, could only belong to God and angels.

I have never taken drugs, though I do smoke, and I don't drink except for an occasional one or two.

By the time they left me, I was so at peace with my life and future that I could hardly wait to continue living. They had promised to come back to help me remember. I know that I subconsciously remembered all of both events, but the messages became confused. They had given me paths to take, and had told me paths for others involved with me.

In the spring of 1985, I was living alone, about six blocks from my mother's house. I often found myself being awakened in the deepest night by a feeling of someone touching me: pushing my stomach; poking my arms and legs; touching my head and neck; what felt like a breast exam and a heaviness across my chest, and someone handling my feet. This seemed to go on for three nights. On the last night, I vaguely saw, in my efficiency apartment, a "little man" running to and around my refrigerator. My door was always locked, as were the two windows.

How long it was after that the next event came about, I'm not sure; maybe it was days or weeks. Then one night I woke up to find myself in a strange room, strapped to a table with my feet up. I felt that my lower half was undressed. I began to fight like a wild animal, with a familiar voice constantly telling me to calm down before I hurt myself. I don't remember them doing anything to make me stop. All I know is that suddenly I could no longer fight. All I could do was cry and babble. At some point, I was told that I was getting a pelvic exam.

I remember waking up the next day angry and feeling dirty, thinking of how "that dream really got to me" and how real it had been.

On another later night, I woke up strapped to a table in a reclining position. The familiar voice was talking, and he informed me that he was sitting at my feet, as I couldn't discover where the voice was coming from. I was too angry to hear everything he talked about, but at some point I began to listen, and then I remember a feeling of deep love for him. At some point I began to beg to see his face. I knew it had to be as beautiful as his words and thoughts and ideals. But he informed me that I'd been looking straight at him the whole time. I couldn't see him. To me, there was a blinding light that surrounded his face. He said that there was no light; he said that I had found him ugly, and that was why I saw the light instead of him. I didn't want this, so my next thought was, "Is he ugly?" He replied back that he thought he was quite beautiful as he knew himself. Also, that as far as his appearance went, it wasn't important, nor did it bother him. Sometime during his words, the light faded and to my dismay and total horror, I saw an unleashed, unchained praying mantis standing in front of me. I was then hit three times between the eyes by a rod. The tip was silver and blue. A burst of pain, each time, lasted only a second. After a few moments, he showed concern for me and my well being, but I began to tell him how ugly he was. He had been right; the horror of what I felt came from what I saw. The beauty I felt from him held more truth.

He continued to talk more. I began to feel horrible about myself and begged him to accept my sincerest apology. He did. I wondered how he could forgive me so easily, because I couldn't forgive myself. He reassured me. He said that our meeting dealt with him and his work. I remember beauty, not of scenery, but of a way of life.

The next meeting seemed to deal with me and my world, as well as my future. There were three paths for me, but it was only during my time with them that I was conscious of which one I'd take. I remember reacting violently to them again, after again realizing that what I was seeing was alien to me and to everyone I knew. I kept exclaiming that they were "devils" and were here to lie to us and trick us. I said they didn't belong here, and that God would kill them. This time I wasn't tied down; I was free. The familiar one seemed to grow angry. He kept saying that I didn't know what I was talking about, and that they were as much a part of the earth as we were. He ended the argument gently, by asking me if I'd truly know an angel if I saw one. I said, "Of course I would!" "No", he said, "only moments ago, before you saw what we looked like, you thought we were angels and you were tranquil. We never called ourselves angels; you did. When you saw us, you became violent and hate filled, so would you know an angel if you saw one?" I still wanted to argue, but he told me that he didn't want to hear anything from me until I knew the truth and accepted it. They ignored me for the duration of what seemed like many, many hours. I watched closely. For the first time, I realized that there were other humans there. I knew some were being helped, at least one or two.

It seemed that the familiar one was sitting there with me, though I wasn't really aware of him. I know this, because at some point he asked if I was ready to apologize. As I recall, my heart, and correctly, my mood had begun to soften as I watched the beings constantly working. I'm a stubborn one, though. He was so close to me, with the huge almond-shaped eyes and triangular face that left everything in my human spirit bare. The privacy of my flesh, bones, and thoughts were gone. In self-defense, I thought that only the devil would do this. I couldn't look at him, but it went beyond him. I was ashamed of my own ugliness. I guess it was my attitude. I wanted to go home.

I began to look for a way out, and saw it across the room. I saw them coming and going and I could smell pine trees, the air of a woods. I checked myself, and saw that I was free. I saw they were all busy, so I snuck down off of the table and started toward an opening I saw. Suddenly I was asked where I thought I was going. "I'm leaving," I replied. "No," he said. "Go sit down." I don't remember the rest of the argument, but I know I said that I wasn't a child, and then he told me to stop acting like one. Before he was finished talking, I was reduced back to a four year old. I remained firm about not sitting down until he promised I wouldn't be tied down. My task was to go back and consider the question he'd asked, "Would you know an angel if you saw one?" I gave no more trouble, though I thought they were being mean to me and I'd never see home again. I saw sadness in the other beings' faces, but the familiar one remained stern. I don't remember how much time went by. I remember giving him an insincere apology once or twice, and was horrified to find that he knew the difference. His disapproval again left me naked in spirit. But he didn't let me suffer this disapproval as long as perhaps he should have. He said, "I believe in you despite yourself. You have the mind to understand and comprehend, if only you want to." His kindness and apparent belief in me were the greatest shame I could receive. I asked him to make me understand, to make me want to, but he said, "No, the decision is yours alone. You are stiff necked; use this trait in a positive way with your choices, and begin to learn."

The next thing I remember is an operating table. It seems that I was semiconscious, and they were at the top of my head. The familiar one was by my side, talking and watching me. He was briefing me on my future. They would do something to my brain to take me to a deeper depth, and then he would drill me again. At first it seemed I remembered well, and repeated things back joyfully. But I hit a point where I recalled only a few words, and then nothing. I was aware of them and him only when he would prod me back to consciousness. But some part of me was always aware, constantly listening, hearing and seeing. I think I vaguely remember what seemed a slight panic, toward the end of this memory. Then the friend said, "Repeat back everything I've said." I was tickled and at peace. "What does it matter, my friend? I am free; now I understand," I said. He looked up and for the first time appeared puzzled. He turned to them and said, "You have four minutes. Any longer, and she can't come back!" What was he talking about? I felt more alive than before. I was energy; there were so many beautiful colors. He looked and acted so concerned. "Sharon, help us!" he demanded. Only they seemed concerned about my body. I felt too good. "You help. I don't want to," I said. He seemed frustrated. "Can't you?"

"Not without you! It is your decision," he said.

I said, "I want to see God; you said He existed."

"He will not honor you if you stop now."

"I am not stopping. I am alive! Look at my body; it is useless," I said.

There was a countdown. "Now! Without you your body is useless, but it is healthy; it is strong!"

I said, "It couldn't fight you and it was helpless, like now. It is nothing."

He said, "This body will give you children, it will feel life, it will know life and it will continue life."

I said that he gave me no choice.

He continued, "For the sake of the children and the honor of God, I'm asking you to let this body continue for the sake of life. Don't quit! "

There is more; I know this. It seems that they left me with a promise. I feel that a great deal of this has been met through my husband and children. I'm presently carrying one more child, I think my last. For me, it's been a long hard road. I wanted to remember them, and it seems they wanted me to, that they have allowed it. I came away from the experience with hope. I seemed to have played the negative and they the positive role, constantly working to show me their points. They shared themselves, and a great deal about themselves.

From the words in your second book, my visitors are the same as yours. In a way, your titles say it all. Both of my counselors fully believed and said that I was sane. The only proof I have is in my mind: the familiar one's voice, the feel of the skin like a "chamois," and their smell of sickly sweet cinnamon/cloves. I remember telling them how bad they smelled. They felt our odor was much worse, to which I retorted that at least we knew it, and used deodorants.

This much I know: Life isn't as black and white as many would like to think. It's colorful beyond any limits, and is a continuance of something far greater.

CHAPTER NINE

THE DEAD

And they kept that saying with themselves, questioning one with another what the rising of the dead should mean.

Mark 9:10

The Visitors and the Dead

These next group of letters illustrate one of the most powerful and important aspects of the close-encounter experience. This is its connection with ghosts and the dead. One of the most moving experiences I have ever had involved a desperate fax from a man whose young son had just had an encounter with his dead older brother, who had appeared in the bedroom surrounded by visitors and said that he was all right.

The older boy had died only a short time before, and the poor family was hungry to know whether or not anything like this had ever happened to anybody else. I was glad to be able to tell them that it had, and to relate the stories of other witnesses such as those who have contributed the powerful statements recorded here.

So what does a thing like this mean? Why would aliens appear in the context of ghosts and apparitions? Probably, because they are not aliens in the simple, conventional manner that is usually assumed. Indeed, the first of these letters began with an encounter in which the visitors not only told the witness that she was one of them, but even included her family name and its meaning! They implied that all mankind may have been split off from some other species in very ancient times, something that more than one witness has reported being told.

To receive letters like these is a deeply humbling experience. When I read of the almost inconceivably painful ordeal that the first author describes as she relates the death of her daughter and the strange, disturbing events that surrounded it, I cried, wishing as never before that the insanity of denial that has afflicted our culture had not prevented her from getting some answers to the devastatingly powerful questions that her story poses.

The second letter is more enigmatic. Apparently, the witness received a visit from her dead father. Although it did not occur in a close-encounter context, she reports that she has been a frequent witness. She was surprised at my reference to a half grapefruit in a speech she heard me make. Oddly enough, so was I! At the time, I thought that the simile I used was oddly inappropriate, and did not know why I'd used it. Was her father calling to her still, through the medium of a speaker whose speech seemed to be getting a bit out of control?

Among the half dozen or so group experiences at my old cabin in upstate new York, there was a continuous link between the visitors and the dead. In one situation, seven people on the ground floor of the cabin were having a conventional close encounter while a couple in the basement saw a friend who had died in an earthquake in Mexico City. In another, our secretary and friend Lorie Barnes was walking down the road in front of the cabin when she came face-to-face with a brother who had disappeared many years before and had long since been declared dead. He told her that he was well, and then dissolved into invisibility before her eyes. That night, the visitors came to the cabin.

The last of these letters involves two experiences, one in New York and one in Los Angeles, that, taken together, also suggest a connection between the world of the dead and the world of the visitors. (Indeed, what if the visitors *are* the dead . . . disguised, perhaps, to avoid revealing their true identity to the living?) The correspondent's first experience is one of the most fascinating I have read, because of all the implications inherent in it. Apparently the soul of somebody who had recently died approached her and *asked* her to have a baby that it could use to reenter life! She did, and the soul continued coming to her even after the baby was born and adopted away from her, only ending the relationship when the baby's growing mind made it impossible for the soul to remain free. Later, the woman had a close-encounter experience with a more conventional entity. In both cases, there was supporting witness.

This group of letters suggests, at the least, that the world of the dead is very close to that of the living, and that it either uses the alien form as a disguise, or is connected to alien life in ways that we are not.

In any case, I believe that these stories and all the others like them are right at the heart of the mystery: If we could understand the visitor experience, I have no doubt that we would also understand ourselves, our true history, the meaning of death, and the destiny of the soul.

A Traveler Returns Home

I was born in Texas in 1945. I've had OBEs since I was a small child, and never thought it unusual, rather that they were private and no one talked about it much. They became very intense during my freshman year in college, leading up to a period where I went completely blind for three days. My ESP seemed to become more and more prevalent, so since I was a "very rational" honor student, I began psychotherapy, thinking something was severely wrong with me and I wanted to get back to "normal" as soon as possible. I got married while still in college and would stay during summer sessions as well. We had a child in 1966. In the early summer of 1968, my daughter and I drove to New Mexico to visit some friends who had a small ranch there, before going west to meet my husband. Leaving Taos, we drove to Flagstaff and waited there until the early evening, to cross the Mojave Desert at night.

About two hours outside of Flagstaff, my daughter began to shout that she saw a spaceship in the sky. How did she know what it was? The sky was unusually cloudy, and I looked to where she was pointing and saw first two and then three lights moving rapidly in the sky, turning at ninety degree angles, pulsating and disappearing, etc. I decided to pull off the road onto a dirt trail I saw to the right, leading into the desert. We were away from the road lights, but I thought I could still see them at a distance. We watched the sky together, she in the backseat and I in front, when suddenly in front of the car there appeared a huge, dark and glowing object with a partial row of lights in the middle.

The next thing I remember is my breath being knocked out of me as I somehow went through the windshield of the car. I remember looking back for an instant and the car was completely empty of myself or my daughter, and I was stepping into an opening in a vehicle. I couldn't see my daughter, and I asked in terror about her. "She's going to be all right," was what I heard in the center of my mind, and I was strangely soothed and unusually happy.

These beings were tall, about six and a half feet, and seemed to be robed in a fabric that emitted a type of light periodically, during movement. Their skin was silvery, and their eyes were round and a violet-blue that sometimes streamed out on me with a feeling of love or long lost family; it was almost like a homecoming. Their eyes were closer to the surface of their faces than humans', and the nose wasn't well defined. Their mouths were fascinating. Sometimes it seemed that they weren't dressed at all, and the body definition wasn't sexually differentiated. I was standing with two of them and noticing that they had no hair, but there was something like fabric that was crumpled and folded behind their backs.

They seemed to be smiling, without moving their mouths. As soon as I thought "hair," one of them seemed to produce beautiful reddish gold hair all over its head. This frightened me. The room I was looking into was about twenty-five feet wide and semicircular. It was rather dark, and filled with TV screens running the full wall area, stacked upon one another three and sometimes four rows high. All sorts of pictures appeared on the screens, and strange symbols, and terrains I'd never seen. Under the screens was a type of built-in desk, curving all along the wall. In the middle of the room was a long table with three or four chairs that were movable. There were three beings in the chairs, two of them facing the screens and moving around, while another one at the desk area stood from time to time, moving things around. They did not look up. They seemed to be of the same slender body type as the two that stood with me, but were not quite so tall. Those two seemed to be laughing all the time and sometimes there was a sound like wind. They kept saying "Welcome, welcome!" in my mind, and laughing. They then told me some strange things about human origins and alien intervention on the planet earth at various times in the past and future. Then they started speaking to me about my individual history. This will sound outrageous, Whitley, but I'll say it anyway:

There was a whole generation of beings that came to earth in the far past and took up earth life. They were from the family of Ranm. That root family name was their name root also, but either that planet wasn't in existence anymore, or it was now inaccessible. They said that was why the old god names were as they were on earth: Rama, Brahma, Raa in Egypt, and Abraham, etc., in order that humans might remember. But so much confusion set in that the names became designations for gods or heroes, and that wasn't the point at all. Rather it indicated the name form of the origin of them, and some of us, being from other star systems. Then they began telling me my name in their tongue: "Shalisha Li Ekimu Ranm," and kept saying it in my head until I got it right. They said those words meant much more, and could be found in earth literature. There was such love flowing through them, as they helped me with the name and the earth lineages that went back to the stars. Understand, this wasn't exactly like words, but were rather images or sound pictures that moved between us.

Then they took me through a gray, curved corridor to the right of the entrance where I'd come in. I can remember not being able to walk, and then walking with ease. We came to a room at the end and to the left of the corridor. This room contained the ship's driving mechanism. This happened in 1968. I was twenty-two and had no idea what I was looking at. In front of me was a huge crystal, perhaps three feet across in the middle. It looked like two pyramids placed base to base, although at times it seemed multifaceted and totally brilliant and jewel-like. The crystal seemed suspended in the air, and around it was a matrix of wires or tubes connected into a solid type of material concealing the ends of the tubes in a dark smooth mass, so that the entire thing rose about four feet from the floor. They told me to put my mind into the crystal, and as I did I'd be able to learn how to fly the ship! One of them telepathed how to do it. I tried and failed, but they kept coaxing me and I could hear them smiling: "Go on, you can do it!" Finally I got it right and we began to move out, first above the earth and then through the angular pattern of space that was also time. I asked why I had to do this, and they only said, "So that you can remember flying and piloting when necessary," and then there was laughter. After the initial information was placed in the crystal and wire matrix, nothing more was necessary, but we stood there anyway until they said, "Time to return to earth."

Frantically, I panicked and asked about my daughter and was soothed again by them saying she was okay. Then they said they were sorry, but didn't say why, and then there was great love. As we moved to the exit place, they said my name again several times, and something about "soul lineage." I was reluctant to go, but the next thing I knew I had gone through the car windshield again, and found myself hanging out of the window gasping for air; I had been crying and was covered with sweat. My daughter was in the backseat crying. She told me never to touch her again, and that she knew who I was and she hated me. I tried to calm her and ask what had happened to her, and she shouted, "I'll never tell you! Leave me alone!" I had a notebook in the car, and before we left I forced myself to write these things down as I remembered them. I looked at the stars in a daze. It was almost midnight, and we had lost about two and a half hours. At that moment the scene seemed uncanny, yet so perfectly normal. I felt then that this was the first time I'd been able to remember, but that it had happened before and I was blocked in remembering. I drove on to California then, as if nothing had happened.

I couldn't tell anyone ever, and swore to myself to never discuss it. I then began getting afraid of going to sleep at night, and became really ill and nauseated. My hair began to fall out and my mouth started bleeding, and I was exhausted. I took more vitamins.

One night my daughter woke up screaming and I went to her bed and she said very factually, "Mommy, I'm going to die. The spaceship people told me so. They said little bugs had gotten into my body and they were sorry, but there was nothing they could do since I'm a little girl." Then she went back to sleep. This frightened me beyond belief. That morning, she woke up with a high fever and had severe joint swelling. I took her to the hospital and she was diagnosed as having rheumatoid arthritis, yet they weren't too sure. They wondered if she'd been exposed to radiation. She was in great pain. I took her out of the hospital and drove back to Texas and put her in the hospital, only to find out that she had a very rare cancer of the nervous system, neuroblastoma, and it had metastasized, and she had just a few months to live. She lived until September of 1969. Before her death, she began to draw extraordinary pictures that were more advanced than a ten year old's, even though she was only three and a half. She began to write poetry, which I sometimes wrote out for her. The doctors were amazed and thought it might be due to the chemotherapy, but were not sure.

The day after her funeral, a friend of mine who was a graduate honor student at the University of Texas called me from Austin hysterically, saying that she had to drive to Houston immediately and tell me something that had happened. Without glasses, she was legally blind, but she drove anyway. I didn't think I could handle another emotional crisis, since I was in such grief, however she came that day. We went for pizza, and she told me what had happened. She said that two nights before at about 2:20 A.M., she was awakened by a noise and then saw her roof begin to dissolve. In the air above, she saw a type of spaceship. Two tall beings appeared, and in between them was my daughter. They told her they hadn't been able to get through to me because of something, but to let me know that my daughter was okay and was with them! She thought at that point that she'd gone completely insane. At that point in her story, I broke my promise to myself and told her what had happened that night in the desert, and we both cried and cried.

My career has changed radically since that year of 1969. I became, due to these experiences, a professional psychic and astrologer and Qabalist. I've never advertised, but have lived and worked all over the world. [Note: She relates that she reached a high position in an eastern esoteric organization, but eventually left it because it was too spiritually confining for her. She asked a spiritual leader to free her of her connection to the visitors, and feels that he may have succeeded.] Mystical experiences continued, but never anything specific about the visitors.

Before that, in 1975, while on an archaeological expedition

in Bimini in the Bahamas, there were two solid weeks of visitor communication. It culminated with a message that they would appear in their vehicle at 9:15 P.M. over at a friend's flat, and that we should all be there. I felt as if I was out on a proverbial limb. Sure enough, though, the craft appeared to all the expedition members present.

There are so many other experiences. I was in Port Aransas when I was twelve, since we spent some of the summer on the water. I have been to England three times in the last year, and many interesting things happened at Glastonbury. There have been probably twenty-five visitor experiences that I have recall of, all of which leave me with an incredible elation for about three days, and then a horrible fear of going to sleep at night. I'll close with what happened in October of 1989, which is what spurred me to write this. A week before, I'd decided that I wanted to regain the connection back to my destiny, although as I look back, that alien connection happened many times, even after [the leader] said he'd take it away. The only difference seemed to be that I wasn't fascinated and fearful somehow, and that it wasn't important. Then, I wasn't interested in anything but the mystery that we humans seem to be part of; I wanted to know that mystery outside of any tradition, system, or dogma, no matter what anyone said to the contrary. On the evening of October 20, I set my meditation circles around the house and bedroom, and went to sleep. I'm no longer married and live alone with three other friends in a fourplex. I wanted to try to get out-of-body, since it was the first anniversary of an intense experience I'd had in Glastonbury, and if I could, I would try to go there.

Sometime in the early A.M., I heard the crackling sound I always associate with out-of-body, and sure enough I could see myself lying in the bed, as I moved in light body through the floor downstairs to see if I could grab my neighbor and get him to have an OBE, too. He saw me, and moved out-of-body, and we roamed around a bit before I departed for Glastonbury. He had always requested that, if I was able to get out-of-body, I should come to get him if I could. This did happen several times, and we shared the occurrences immediately in the morning.

Usually I was drinking coffee, and he'd knock on the door at about 8:00 A.M. with, "You'll never believe what happened to me last night." Usually I said, "Try me." The recalls were exact.

After an extraordinary experience at Glastonbury, I came back in through the roof and sat up in my body. It was about 4:00 A.M. Then I went back to sleep, with no thoughts of visitors. I was awakened awhile later by two taps on the glass window in the bedroom that have a small, slanted roof under them. Since I live upstairs, this is the only area vulnerable to entry, and I keep the windows and screens locked. For the tap to happen, the screens would have to be pulled open. As I awoke, there were two more raps, one long and one short. I turned over and looked toward the window. The moon must have moved to that area of the sky, since there was a lot of light streaming in and casting moving shadows of leaves and branches on the half open blinds. There's also a light that comes in at night around the corner, attached to my neighbor's wall, as well as a nightlight that we had installed to be activated by body movement. It seemed they were all on. Staring in at me was a little gray being such as you described, but the cranium was larger. I could see it through the half open blinds, and also its shadow, which unlike the branches stayed perfectly still. My first thought was, "That's the being the guy who wrote that book was talking about." Then to be sure I was awake, I looked at my hands and sat up in bed and pulled the covers down a bit. If I'm out-of-body, these actions are impossible; either I can't move my body in the physical, or my hand goes through matter and has no effect on it whatsoever. In those circumstances, only thought causes matter to move, nothing else can. I sat up and looked at it, and it looked at me. It didn't enter my room, perhaps because of the magical circle I had previously set. Nothing else I know of could have kept it out. We stared.

"You don't have the right to knock on my window and wake me and come to me with your mind, uninvited, and expect me to go with you. Come back when we've both agreed to a visit, not like this at 5:15 A.M.!" I felt absolutely no emotion from the being surveying me. Then I turned on the light. "Go away!" I said out loud. "You're scaring me." There was no soothing energy or anything. Maybe it was drawn to me because I'd passed through its world while out-of-body, and it had come to see me in my world to show me what it was like; I don't know. I looked at the digital clock, and it was 5:20 A.M. About then, there were noises downstairs. It sounded as if my neighbor had begun to open and slam his windows shut, and then the closet doors. There was also a pounding sound, like a few people running hard on the floor downstairs. In the night silence, it seemed loud. The windows continued to slam, and I thought he was participating in some sort of bizarre behavior, and made a note to inquire of him in the morning. With the light on in the bedroom, I could not see outside. I reached under the bed for my revolver, and mentally told the figure that if it didn't leave, I was going to shoot at it. If it came inside, I'd shoot it for sure. It sounded like my Texas upbringing talking. Thinking back on the incident, I wouldn't have shot it, but I can never know for sure. When I turned off the light it was 5:30. The being had moved to the other window by the slanted roof. It seemed to be sitting on the roof and looking in. I laid back down with the gun in my hand, on top of the covers, and stared at it until I went to sleep. The noises downstairs had finally stopped before I dozed off, and I felt strangely comfortable about the whole scary thing.

The next morning at 7:30, I sprang out of bed and tapped on the window. Yes, that was the exact sound. I went downstairs at 8:00 and was looking up at the roof when my neighbor came out. It had been another interesting night, he said, and guess what? I'd come to get him again, but this time he was able to come upstairs to my place too, before we had gone into that golden twilight place we always went, where I showed him how to go through walls. I laughed and said I remembered, but not exactly in the same way he did, since I had been able to successfully go to Glastonbury in what seemed like an instant. Then I told him what I'd seen on the roof, looking in the window. They could have climbed up there from the porch without much trouble. Then I asked him about all the noise downstairs. He hadn't heard a thing. He'd had a dream that was disturbing, however, after the OBE. He said that there seemed to be someone up to no good who was trying to break in downstairs, and he was able to keep them out of the house. They then went next door in a vacated area of a garage apartment and began drawing up plans for the next attack. They were up to no good, he reemphasized. He was doubtful about the experience and said he thought that I believed it, but a real physical being? It had cast a shadow, unlike a projection, and I'd been wide awake sitting up in bed. I woke up with the gun on the top covers. It was real.

THE COMMUNION LETTERS

"This Is Not Happening"

As I have written you in the past, I, too, have had experiences with aliens or beings from somewhere else. My experiences began approximately in 1983 but stopped around 1993.

During your talk in Dallas [January 1996] about people visiting your cabin in New York you made reference to "half a grapefruit." My husband heard you and so did the lady sitting next to me. It has really made me curious and after I explain I think you will understand why.

Early on a Sunday afternoon in about 1986, while home alone, I had just finished some weekend chores when I went into the living room to relax with a book and wait for my husband to come home from work. As I walked into the living room, I heard an odd noise, so I looked out the front window and saw a 1940s gray pickup truck pull up to the curb out front. My father died in 1969, but I watched as he got out of the truck and removed metal baskets filled with fruit from cabinets in the back of the truck. I watched him cross my lawn, but thought that couldn't be my daddy as the man was much younger. By the time the doorbell rang, I was in a mental turmoil. I peered through the peek hole and thought, "This is not happening," as he peered back at me. On opening the door, I could barely manage the word "Yes?" He was even dressed as my father would have been on a warm summer day. In a clear, resonant voice my father threw back his head and began describing this wonderful fruit. He smiled just like my father as he put the baskets down and took out an orange in one hand and a grapefruit in the other. I was experiencing all kinds of emotions during his presentation. I even had an overwhelming wave of pity for the poor peddler as I started to unlock the screen door. Before I could step out onto the porch he reached into his pocket and whipped out a switchblade knife and sliced the grapefruit in half. I could barely manage to say, "I'm sorry, I'm not interested."

I've never seen a more disappointed look on a person's face as he placed half the grapefruit on the railing of the porch, picked up the baskets and went back to his truck. I watched from the front window as he drove off.

I wondered why he didn't go to any other houses and called a neighbor who said she had not seen a peddler. I went out on the porch to discard the grapefruit, but found I could not touch it, and later had my husband throw it in the outside trash. This man was probably in his late thirties. My father was forty-one when I was born and at one time had been a Raleigh salesman in east Texas, but I never considered him a peddler. He even had a gray pickup truck. It's been almost ten years since that happened, but I think of it every day. I know that it was my father, but I was too scared to talk to him and even said I wasn't interested and hurt his feelings. Isn't that sad?

You may now understand why I wonder why you referred to half a grapefruit.

"GHOSTS"

We live in a very old home that dates back to before the Civil War. It is complete with cellars, a sealed-off attic, and a huge graveyard just out the back. Until I read your book, my family and I believed that what we had here were a bunch of good old-fashioned ghosts.

One of our visitors is the "Sentinel," who checks out the bedrooms every night. We also have a tall, blond "Guardian Angel." All the others are very small and avoid being seen, except on rare occasions when they show off to strangers. The gray forms that you speak of are forever darting around. They seem to be about four feet tall. There are also white, shapeless things that stay primarily on the second level, which is the living area of the house. These are the pranksters who scare our guests.

We also think there's a bad entity here, that we've named "The Critter." He comes out of the cellars occasionally to terrorize the household with his howls, growls, snarls, and slow-shuffling footsteps. He is big and black and wears a hooded robe. There is an urn hanging from his side. He was first seen by my sisters on October 17, 1939, a few seconds after I was born.

The Guardian Angel is the only one who has communicated with us. This usually happens when I'm in a dream state. About three months ago, my oldest son came home late, when the rest of us were asleep. He had been drinking and was tired, but he decided to fry some potatoes anyway. He put them on the stove, turned the burner up high, then fell on the couch and went to sleep. The frying pan and stove caught fire.

Upstairs, someone shook me violently, calling out, "Wake up! Get up!" The whole house was filled with smoke. I woke up my husband and we ran downstairs to put out the fire. The kitchen was ruined, but we could have all been dead. The next day, when we all had our wits about us, my family asked me how I managed to wake up in time. When I told them, my daughter remarked, "You bruise so easily; if that story is true, your arms must be a mess." Sure enough, I looked and saw horrible bruises on both arms and shoulders, as if large hands had shaken me.

We can all recall dozens of other incidents, both good and bad. My family tells me that I should mention the different odors that are associated with our "friends." One smells like rotted earth with a touch of sulfur; another smells like cabbage cooking, and others smell like perfume and Old Spice shaving lotion. My husband gets particularly aggravated with the Prankster, who twists his toes in the middle of the night. Other than that, we live in harmony with them.

From Soul to Son

As far as I know, during this lifetime I have not had any direct encounters with the visitors. However, I have had various experiences that defy our currently accepted reality.

1966—*New York City*

I was living in an apartment shared by one other person. She and I had twin beds with hers nearest the doorway (located next to the head of her bed) into the living room. We had both just retired for the evening, neither one of us yet asleep, when I sensed someone entering the room then stopping at the foot of her bed. Although I did not see him with my body's eyes, I knew it was a young man about six feet tall, that he was very sad and confused, and that he was there to see me.

I asked my roommate if she had just noticed anything. She said yes and, without telling her what I had just sensed, I asked her to tell me what it was. She reported exactly to me what I had just experienced. During her telling me, the being moved over to the side of my bed and looked down at me. I felt no fear, only curiosity about why he was there. I tried to determine whether I felt that I knew this person—was it an uncle who had died some time ago or my grandfather who had also passed away?—but I did not really feel as though I recognized this being. From what I gathered from his thoughts, he had recently been killed in some sort of accident and was disoriented, but knew that he needed to get back into a body. He basically was asking to let him have a baby body from me. This threw me into quite a turmoil. On one hand, I felt this was perfectly okay and could be an interesting experience for me (I was twenty-one and had not been married yet); on the other hand, I thought—what would my parents think, what would their neighbors think, etc. Ultimately, (which occurred probably in only a matter of seconds) I felt myself agree.

It was not until sometime after this experience that I discovered I was indeed pregnant, which I assumed was from the boyfriend I had split up with not long before. I went off to have the baby in secret (and to then give it up for adoption).

In the meantime, the being seemed to have dropped the mock-up of the body that my roommate and I had originally sensed and became a ball of light that mostly hovered in one corner of our living room. This is where people would sense his location. This being and I continued telepathic communication for approximately four years when, finally, I sensed a disturbance coming from him which seemed to be that this connection was now becoming a confusion in his new life. I let him know I understood and we ceased having the communication.

1971—*Los Angeles*

A friend and I had decided to stay up all night in an attempt to complete some studying we were doing. We were working away intensely, hoping to finish up as soon as possible. At around four o'clock in the morning, I looked over at her sitting in her chair when I saw a little being about two and a half to three feet tall appear right beside her, Although I could not make out any details, the best way I could describe it is as though it were in a kind of grayish translucent cloak of light. It was only there for a moment when I saw an extension come out of it, like a pointed finger, that pushed my friend. At that very instant, she fell off her chair in the direction in which she had been pushed! After this, we both started laughing nonstop. She had not seen anything, but she told me she had felt something push against her shoulder that made her fall off the chair. Neither one of us experienced fear. We both found it very funny and I felt as if the being was just having fun with us, probably thinking how silly we were staying up all night like that.

Message from a Dead Brother

I'm twenty-two years old, and was born and raised in a suburb of a big city. After nineteen years I met my father, and then learned that I had a sixteen-year-old brother who was killed in a hit-and-run accident. His name was Donnie.

I'm writing you with the hope you can shed some light on a situation I feel is driving me crazy. I'll start with a dream I had three years ago:

I was sitting on the sofa at my dad's at Christmas time. I fell asleep, and suddenly the Christmas tree behind me gave off a blinding white light. My brother Donnie walked out of this light and told me to be sure not to wake my dad, and to follow him. He led me to the back door, and as soon as we stepped out, we were in a wooded area. As we walked down a path through the woods, Donnie told me not to blame myself for not knowing him, that he was always there when I needed him. He talked more, but I'm not clear about what he said. The last thing he said was that there was something he had to show me. He led me to a tree stump with a blue photo album sitting on it, and then he vanished. When I opened it, I found numerous pages with pictures of UFOs. The disk-shaped objects were brown, and seemed to be bilevel, with a row of windows on their top halves and a row on their bottom halves that looked to be tinted black. The pictures all seemed to be of this same disk. Some of them were up in the air and some were on the ground with the door open and a light shining out of it. All were close-up shots.

I can't even begin to explain how real this was; I even have trouble calling it a dream. When I awoke, my body was cold and I could still smell the night air, as if I had been outside. What concerns me is that this was more than a dream, because when I described it to my dad he unpacked one of Donnie's boxes and showed me the same clothes he had worn in my "dream."

I was fine for some time after that, then suddenly I became paranoid if I was left alone for any period of time. Even in my apartment, I would not enter a dark room without reaching around the corner to turn on the light first, and I constantly felt like I was being watched from around a corner. My girlfriend became concerned. She said I looked like I'd seen a ghost when a car's headlights hit our living room window one night. I remember my reaction, but didn't realize she'd noticed until she mentioned it a week later.

About a month ago I became fascinated with UFOs. I went to the library and checked out every book they had on this subject. That is how I discovered *Communion*. You mention a little white ghostlike figure. One night when I was a young child, I was in the state of being not quite awake or asleep, when a white hand came up from the end of my bed and grabbed my foot. I shook it loose, and when I opened my eyes, I saw the hand go back down.

FAMILY IN THE SKY

My experience began when I was five years old. I had just been taken away from my biological mother and placed in a foster home. My foster parents lived at the foot of a mountain. The year was 1959. The memory is like a piece of film footage, and it never changes.

I am sitting in the backseat of the car, with my foster parents up front. It's night, and we are going home, traveling on a gravel road toward our house. I'm looking out the window at the stars. The car is very quiet inside, as if my parents are not talking. There is a weird stillness, and then something flies over the top of the car. I think it is a jet, because it has flames shooting out its back.

It seems that the paranormal has been part of my life, for most of my life. I was constantly leaving my body as a small child.

My greatest experience was when I was walking home late one night from a friend's house. This was a seven-mile walk along a river and across a bridge. As I crossed, I noticed something strange. I was glowing; everything was glowing. What looked like a glowing fishnet crossed the entire sky. To my left, upriver and above at about two thousand feet, something that looked like a giant manta ray without a tail sat motionless. It was huge, four times the size of an aircraft carrier. I said to myself, "It's a ship!" I sat on the riverbank and watched as the sun came up. I could still see the lines in the sky, but they faded around noon.

Strange things then began to happen. What appeared to be a blue glowing ball with sparks flying off of it hovered at eye level, just barely out of my left field of vision, keeping a distance of about a hundred feet from me. Walking down a street at noon, I passed a pretty girl dressed like a gypsy. She looked right at me and smiled. We were going in opposite directions. I walked another block, stopped and looked again, and there she was looking at me. I know she hadn't turned around to come back. That scared the crap out of me. Several more times, people who didn't seem to fit into the local culture appeared, always looking at me as if they knew me. It was hard, because I was alone with these experiences, and had no one to talk to. When I did try, my friends avoided me for a month or more.

By 1984, the gates of heaven and hell opened wide. One cold winter night I walked outside and looked straight up into a starry sky. Suddenly it was as if my mind exploded into a billion pieces, as if the universe was inside me, and I was everywhere inside it. I experienced omnipresence. This is really better than sex. Then I experienced the dark side.

One night I heard something in the hallway of our apartment. I got out of bed, turned the hallway light on, and saw a creature. It looked like a skeleton covered with rotting flesh, flying down the hall with a flaming sword over its head. There was a howling sound, like a wind, and it was over in three seconds. The next thing I knew, I was on the kitchen floor, shaking and sobbing. I looked up to see my live-in girlfriend trying to help me stand up. There was genuine fear in her eyes.

In 1987 I moved. My paranormal experiences continued. Within the first six months, I had a dream in which I met my deceased biological father. He told me to pursue my writing. He said I would have something important to tell people in the near future. When I awoke, I opened the nightstand drawer next to my bed, for no apparent reason. I was stunned to see my birth certificate lying on top of my other papers, when it had been tucked away in the bottom of the drawer the night before. On top of my other dresser was a pile of pennies, except now they were neatly arranged in a perfect circle. I shrugged all this off; it was not impressive compared to other experiences I've had.

In 1989, I was living in California. On a cold Saturday morning in October, I saw an amazing sight. In broad daylight, I saw an oblong, egg-shaped ship, maybe thirty feet long, moving slowly up the canal. It was easily thirty feet above the water. It moved at about eight miles per hour, and acted as if it didn't have a care in the world. It was blue, and had oblong portholes down its middle, from front to back. This was never reported. It was as if I was the only person who saw it, in broad daylight in a populated area. That was frustrating; everyone thought I was crazy.

In 1992, I think I screwed up. I was living in the heart of a downtown area. I awoke at about 3:00 A.M. to see one of your little visitor friends crawling through my window. I don't know why, but I said, "Not yet—I'm not ready." The little guy just turned around and left, and I fell back to sleep.

I had just broken up with the woman I thought was the love of my life. The breakup was devastating; psychologically I was a vacuum. For the next three years I lived in isolation, and spent every day reading every book I could on philosophy, esoteric knowledge, and religion, literally hundreds of texts.

Last year I had an amazing dream. I was on a ship with people who didn't quite look like people, about twenty of them. They seemed overjoyed to see me. One said, "You've been gone so long. It's good to see you again." I felt as if I'd come out of a state of amnesia, as though this was a place I'd originated from.

I awoke then and the memory faded, but not completely. When I think now about the dream ship and those people, I feel overwhelming love and a deep connection, like family.

AFTERWORD

And so, its silks and jewels and mysteries piled high, trailing exotic scents and memories, the caravan passes by. But we are left behind, all of us, even the authors of the letters.

The problem is, nobody knows what kind of experience they reflect, let alone what they mean. There is a tendency to pigeonhole. Scientists of the brain are tempted to dismiss them as the output of temporal lobe epilepsy or bipolar disorder or one of the more gaudy psychoses.

But, if this is the case, how can their extensive multiple-witness content be explained?

Presently, all a scientific observer can do to maintain the integrity of theoretical assumptions that suggest that these narratives are the trivial byproducts of known disorders is to dismiss the parts of the stories that don't fit. However, that isn't science, it's fantasy.

These stories are probably too strange to have emerged on their own out of even extremely distorted imaginations. I have read much material written by schizophrenics, paranoids, temporal lobe epileptics, and others, in an effort to compare it to the letters.

While some of the abnormal material is indeed profoundly bizarre, it has a disorganized quality and a lack of coherent context that sharply distinguishes it from these letters. There is absolutely no consistent symbolic or structural content across the body of narratives produced by the mentally ill—and this makes these letters radically different from that product.

Although by no means universally present in the letters, probably the most telling symbolic consistencies involve owls, wolves, and ritualized knocking sounds.

These symbols are not part of our culture. In fact, they come to modern minds like words from a lost language.

But they are not arbitrary, far from it. Once, they held rich meaning. To the Greeks, the owl was sacred to Athene, the ancient goddess whose staring owl-eyes also glare out at us from even more ancient Sumerian images. As well, she was the goddess of Mari, the mother-city, who gave her name to Mary, the mother of God.

Christian legend has it that the owl was a banished demoness, condemned never to look at the sun; earlier, the owl represented wisdom.

Whatever they really are, the staring eyes of the distant past that belong to the "eye-goddesses" Ishtar and Lilith, and all their sisters, seem to be appearing again, stunning close encounter witnesses all over the world with their fearsome demand that we face the unknown.

As for the image of the wolf, it was a she-wolf who suckled Romulus and Remus according to the Roman legend, and thus it was she who was responsible for the creation of the city upon whose shoulders Western civilization was founded. For it is Greek ideals that animate our culture, but the Roman legal structure that makes it run, and Roman Christianity which was the wellspring of its ethics.

Knocking has been a feature of spiritualist experience since the nineteenth century, almost as if, in those years, another world began knocking on our door.

In Masonry and in many other secret traditions, knocking symbolizes the passage from one level to the next, and the knocking on the door is said to awaken the soul.

Thus, there is a surprisingly clear message being communicated, although using symbols that have long ago been dropped from our lives.

Even more consistent than the symbolic content of the letters is their strangeness, which is so extraordinary that it seems almost to place these narratives beyond the scope of the human imagination. The aliens, if that is what they are, don't act like us. They don't even act like our science-fictional characters. But they *do* relate to us, using symbols from the wrong era, perhaps, and generating impressions that are profoundly discordant—but also coherent, for they turn out also to be profoundly meaningful.

Beyond the degree of symbolic consistency present here, there is also a consistent structure that emerges out of the strangeness itself: the witnesses are almost always called in some way. They are addressed by shadowy figures, by lights in the night, by touch, by whispered words, by thundering low voices, sometimes by the cries of their own children.

However it emerges in an individual situation, this call has a very particular effect, for what it does is to cause the witness to end up facing an urgent question that can neither be answered nor ignored.

In the past, we have always closed these questions in the same way: with superstition and blind belief. Our whole mythology, in some ways, can be seen as an attempt to draw meaning from experiences like these. The modern UFO stories may be closer to describing the reality behind such reports as these than were the stories of elves and gods and angels, but how can we know?

Right now, the absence of serious scientific study means that this vibrant, immensely provocative material has the same potency that was possessed by the mythologies of the past. What we have here, at the moment, is the bone-structure of another new religion.

But must we leave it at that? Hasn't the past taught us anything at all?

There are so many new discoveries taking place right now about the nature of the universe, the role of life in it, and the secret structures that govern the relationship between the observer and what is being perceived, that it would seem that a completely new approach to this sort of material is potentially available, and with it a means of deriving some useful meaning from it at last.

By applying modern testing and diagnostic tools and

appropriate methodologies to this material, we can make a break from the past, for the first time addressing this phenomenon in an objective manner.

This gets to what I think is the core opportunity: We have a chance to ask some new questions, and ask them with techniques that can lead us to some real answers.

Light, in other words, can be shed into this remarkable darkness. We should begin, perhaps, by simply dropping the debate about whether or not aliens are here. This debate is the equivalent of the Middle Ages' discourse about the number of angels who can stand on the head of a pin. It is pointless, futile, and a waste of resources. Until we have some facts, who knows what's really happening?

Looking objectively at the enormous question implied by these letters and the vast ocean of similar narratives that stand behind this small sample, can only add to the content of science. Whether we discover that an alien presence led to these narratives or not, we will certainly discover things about ourselves and the way we have in the past generated the content of culture that will vastly improve our understanding of human nature and experience.

At present—and at the very least—all we can do is enjoy the mystery of these stories, and wonder at what must have come out of the night or the secret reaches of the mind to disturb the lives of these witnesses.

What, indeed?

WHITLEY STRIEBER is the author of many novels and works of nonfiction, including such legendary titles as *The Wolfen*, *The Hunger*, and *Communion*.

ANNE STRIEBER is also an author and has been involved from the beginning with the answering and organizing of the correspondence that make up *The Communion Letters*.

Whitley and Anne live quietly in San Antonio, Texas, with their son and several cats.